BS 7750

BS 7750

Implementing the Environment Management Standard and the EC Eco-Management Scheme

BRIAN ROTHERY

Gower

Published by
Gower Press
Gower House
Ashgate Publishing Group
Croft Road
Aldershot
Hampshire
GU11 3HR
UK

Gower Press
Old Post Road
Brookfield
Vermont 05036
USA

British Library Cataloguing in Publication Data

Rothery, Brian
  BS 7750: Implementing the Environment
  Management Standard
  I. Title
  658.4

  ISBN 0-566-07392-7

Typeset by Keyboard Services, Luton
Printed in England by Clays Ltd, St Ives plc

Contents

v

CONTENTS

CONTENTS

Preface

This book is intended for manufacturing and service companies who wish to implement an environmental management system which will fully meet health and safety regulations, local authority and national environmental requirements, and the latest industry codes of practice, and which will also give legal protection against claims under product liability and against unjustified charges of negligence.

It is directed at senior management, including chief executives, quality managers, and all executives in charge of operational processes and transport and distribution networks. The system proposed here is intended to meet both the requirements of BS 7750, as described in the BSI published document *BS 7750 Environmental Management Systems*, and the EC Eco-Management regulation. It will also meet the requirements of the compulsory health and safety regulations, and secure third-party accreditation of safe and environmentally good practices, which may provide valuable legal protection, both corporate and individual.

As this book goes to press, two important developments are under way. Firstly, a BSI scheme involving several hundred companies in a pilot implementation of the standard has not yet been completed. When it is completed the findings may result in some changes to the standard. Secondly, the EC Eco-Management and Auditing Scheme has just been passed by the European Parliament, but has not yet been adopted by the European Council. Neither of these two important developments should change the basic elements in the approach advocated here to implementing the standard. Indeed, it is the hope of this author that the book will complement the work of the architects of the standard and assist companies in its implementation.

The text is organized in a manner which will allow both the project to be implemented and the resulting system to be maintained. Of particular value should be the sample documentation, especially the registers and environment management manual featured in the Appendices.

Brian Rothery
March 1993

PART I

The Context

1

The New Environment for Business

There are profound changes taking place in the world of business. These cannot be separated from global social trends which are changing the way we think about and treat the world we live in. Both manufacturing and service industries have come under immense pressures to lead the way in helping to create a more caring operational environment. Market, ethical and legislative pressures are demanding operating standards which are supported by independent third party accreditation.

Manufacturing industry felt the first effects of these pressures, but they quickly moved out to the retail and service sectors, as buyers became aware of the new standards and those standards themselves created new expectations.

The single most important ingredient which has been identified within industry and extracted for rigid analysis and control is the quality of the operational process. In the first instance this was powerfully assisted by

ISO 9000, the international quality management standard, based on BS 5750, but the ink was barely dry on the ISO quality standard when the concept of 'quality' itself was enhanced to embrace operational effects on the environment. This is particularly seen in the latest drafts of the ISO 9000 standard where the authors include health and safety, environment and public safety under 'quality'.

ISO 9000 was itself a revolutionary new standard, the first international standard applied to a system of quality management, that is the whole manufacturing or service process, as distinct from a standard for a product or process. The new BS 7750, the Environmental Management Standard, completed the revolution, becoming what some are calling the 'ultimate standard', and creating a state of quality greater than the sum of two parts of quality management and environmental management. This is because the two, together with independent third party accreditation, make truly facilitative industry a reality, that is an Industry that is both quality-managed from the product or service point of view and that is green and clean, and demonstrably so.

This is quite revolutionary and it both empowers and enshrines industry in its fundamental role of maintaining society, within an environment which must also be maintained.

1.1 Who will be affected by the standard?

BS 7750 will follow the course of ISO 9000 and affect manufacturing industry first. Even as the standard was published in April 1992, the demands for demonstration of environmental consideration were increasing on manufacturers. Many service industries such as transport, power and mass retail outlets using packaging were also under pressure. The standard will eventually reach all sectors of manufacturing and services.

To answer the question of who initially will be affected by the standard is simple, because that answer is all those companies currently affected by environmental legislation and regulations, whether they concern air, water, waste, noise, product use, safety, or materials employed. The new standard will help such companies to control their operations, maintain them within the regulations and demonstrate conformance with those regulations.

There is a large disparity in both the need for the standard and the degree of difficulty in implementing it. High on the list of those who both need it and will find it difficult to implement are those industries which traditionally have to deal with environmentally difficult products and processes, for example, coal and oil-fired power stations, large food processors, and mining enterprises. Ironically, chemical plants, with a great potential for destructive environmental occurrences, may find implementation easier as their processes are inherently controlled.

There are also industries which will find it easier to be first in their sectors with the standard, such as large software distributors whose issues will be product use and packaging, distilleries, natural gas, electrical and water utilities, and, of course, telecommunications companies.

The demands for the standard come from the same direction as those for ISO 9000. Firstly, environmentally conscious buyers, both government and private, demand it, followed by large High Street retailers, who want a green image. Manufacturers will, by necessity, have to force it downwards onto their suppliers. Once the first companies become certified and start to promote their premier status, others will be forced for marketing reasons to follow. This has been the pattern for ISO 9000 and it will be the pattern for BS 7750.

1.2 What is involved?

Anyone who has installed ISO 9000 will be comforted to hear that the approach is similar and that ISO 9000 puts one well on the road to BS 7750, which point we discuss in some detail later. For newcomers to both standards, one can seek to attain BS 7750 in conjunction with ISO 9000 or as a separate goal. At the time of writing there is still no ISO number for the environmental standard, but the direction appeared to be that the environment standard would encompass both quality and the environment, and ISO 9000 itself has started to take on this role.

There are two parts to the answer to the question of what is involved. Where environmental considerations are simple, the job is simply to document and implement a system which reliably manages those considerations and can be demonstrated to a third party accreditation agency. Where environmental considerations are considerable, the biggest job may be to identify and document those considerations and then implement control procedures. These issues are dealt with in the chapters which follow.

In general, what is involved is a process of identification of the corporate need for the standard, a commitment to go for it, an awareness of the specific environmental considerations of each company, and a management system with controls and documentation. Chapter 13 covers the need to build an inventory of environmental directives, but here we can say that the achievement of the standard will be dependent upon the ability of individual companies to demonstrate their knowledge of the relevant directives or regulations which apply to them and their success in implementing a system to cater for them along the lines of the standard.

1.3 Marketing, public relations and legal implications

The marketing and public relations implications of the standard are enormous. ISO 9000 has already given a severe jolt to the public relations profession, but much more is in store as a result of the environment standard. What ISO 9000 did was to eliminate all spurious PR-motivated claims for 'quality' and replace such claims with real, demonstrable quality. This meant that real service began to compete vigorously with unreal PR claims for service. In other words, an actual, good-quality product or good-quality service made a certain amount of PR redundant, and PR, not based on real quality or service, was in danger of rebounding onto its exponents.

Great as the demands for quality are, those for environmental care are even greater, and indeed may embrace quality, for one cannot produce good 'quality' products at the expense of the environment. Just as the word 'quality' was abused before the emergence of ISO 9000, so we have seen, and still see daily, the abuse of such words as 'green' and 'clean'. Real green and clean products and services will meet certification to BS 7750, so the marketing and PR implications are very clear indeed. Companies will compete vigorously to be first on the market with the new verifiable claims.

The literature accompanying the new standard is clear that it expects large marketing implications to surround the standard. Its architects are certainly aware of how the hype surrounding ISO 9000 help promote it worldwide. What we are actually seeing here is a global facilitative trend, which in turn is based on a dominant tendency which propagates itself, and grows almost biologically. A sociologist or biologist looking at the two standards could be forgiven for believing that they were both necessary evolutions in the course of mankind's progress towards a better world, one where industry must continue to be at its heart.

Later we look at the different motives for adopting the standard, some inspired by compulsory regulations, some by virtually mandatory marketing demands, others by a desire to be seen to be 'clean and green'. One important market expected to demand environmental assurances is the recently opened-up EC public procurement market — the enormous market for the purchases of all EC public bodies. This embraces public works such as roads and other construction projects, all supplies of materials and components and all services, above certain threshold limits. Billions of ECUs formerly spent within national boundaries will now be open to competition from companies in other member states. Registration to ISO 9000 has been one formal requirement for access to this business (when the price was also right) but now we can expect demands for environmental accreditation, agreed by BS 7750, to join the demands for accredited quality. The red tape in achieving registration to ISO 9000 has been eliminated to a large extent by each member state setting up its own certifying agency or agencies, which, in turn, accredit to the agreed 'harmonized' standard.

Within the EC, standards and the regulations which may accompany them are not so much intended to be strict legal specifications than support for laws to protect public interest, health, safety, environment and to facilitate trade within the Community.

The regulations are usually called EC directives and they employ the principle of reference to an EC standard, for example, EN 29000 based on ISO 9000, in turn derived from the original BS 5750. In this respect the Commission has devolved full authority to CEN members (see Section 1.6 below for explanation of CEN) who have voluntarily agreed that European standards once approved must be adopted nationally. We must bear in mind that CEN members are already using international standards such as ISOs. The actual legal significance of this for manufacturers is that product conformity to European standards entails conformity to the legal requirements of EC directives. This would be relevant in any European court, and in charges of product liability.

Finally, on the matter of legal considerations we must return to one important issue and raise another. The health and safety regulations are both legally mandatory, from the state inspection point of view, and a corporate necessity for the avoidance of claims for damages from staff. The environment standard, however, has also highlighted two further areas of possible danger. Firstly product use, including its ergonomics, has become an environmental and therefore potential legal issue, and we will return to this fundamentally important point later; secondly, operational safety, in particular controls relating to accidents and emergencies in certain industries, is now so clearly specified as to leave management open to prosecution for criminal negligence should it be not adequately covered. The standard is clearly a way to demonstrate legal compliance with all environment and health and safety demands.

1.4 The expansion of quality standards

The proliferation of quality standards has been quite phenomenal. In Chapter 2, on the background to the standard, we will look in more detail at the evolution of BS 7750, and at its beginnings, but ISO 9000, based on the sister BS 5750 standard, has, at the time of writing, begun to sweep the world.

The adoption by the EC of ISO 9000 as a harmonized standard for the Single Market gave the first real impetus to the notion of using an accredited quality standard to replace unverifiable claims of quality. The moment the standard became the norm within the EC, the inevitable international reaction followed. First US and foreign-owned multinational companies manufacturing in Europe adopted it. This in turn led to a two-way spread: firstly, horizontally into the sister companies of the multinationals and into the parent, in the US or wherever; secondly, vertically, mainly downwards to suppliers, but sometimes upwards to customers.

Once it became apparent that Europe was adopting a standard, which could become a barrier to entry to companies outside Europe, even though it was an international standard, the process began to spread around the world. Growth was aided by the existence of national standards promotion and certification bodies, and hampered by their absence.

Fundamental to the acceptance and growth of the standards, both quality and environment, in Europe, is that accreditation in the country of origin means acceptance throughout all member states. It is in breach of the Single European Act and its subsidiary directives to demand a local standard where there is an agreed European standard covered by a European norm, known as an EN. Thus ISO 9000 is equivalent to EN 29000. BS 7750 will form the basis for meeting the new EC environmental directive or Eco-audit and will also result in both an ISO standard and an EN.

These processes will propel BS 7750 outwards globally, and vertically and horizontally; out to non-EC based companies, downwards to suppliers and upwards to customers, repeating the epidemic spread of ISO 9000.

1.5 EC, international and national standards and regulations

It may be useful to point out the differences between standards and regulations and to relate national and EC standards to international standards. The EC publishes directives which member states may make compulsory with legislation such as ministerial regulations or Acts of parliament. The health and safety directives issued by the EC have become compulsory regulations within member states. Somewhat confusingly the EC may also publish regulations binding on member states, for example, environmental regulatory and accreditation schemes. A national standard, such as BS 7750, which becomes an internationally accepted standard, or one already so accepted, such as ISO 9000, is a mechanism for meeting specified levels of performance and whole sets of regulations, and also of demonstrating conformance to both.

Through ISO, the international standards organization, national standards become harmonized and made 'standard' worldwide. ISO 9000 was the supreme example, before BS 7750, of a single universally accepted quality management standard, for a system as distinct from a product or process.

1.6 CEN, ISO and national standards bodies

CEN is a European Standards Committee. CENELEC is the European Electrotechnical Standards Committee. CEN and CENELEC thus represent both general standards and the electrical standards, which perhaps were the first formalized standards, preceding international standards bodies.

The members of CEN are the national standards bodies of each EC and EFTA (European Free Trade Association) country, *viz*, BSI (UK), DIN (Germany), AFNOR (France), NSAI (Ireland). The members of CENELEC are the Electrotechnical Committees of each EC and EFTA country, *viz*,

BEC (UK), DKE (Germany), ETCI (Ireland).

CEN/CENELEC form the Joint European Standards Institute on common matters and in particular provide the EC Commission DG III (Internal Market) with a single European body, separate from governments, to provide European technical standards (called EN) for publication as *harmonized national standards* within each member state. EFTA members have, of course, also agreed to harmonize, to retain a unified system. When the EC came into being, it found itself with an existing highly developed, unifying infrastructure for harmonizing standards in CEN/CENELEC.

In 1987 the EC Commission requested CEN/CENELEC to adopt the international ISO 9000 series of standards as the appropriate European Standards known as EN 29000. This was a decision of momentous importance for industry worldwide and for the creation of the Single Market. Now another ISO standard, or an expansion of ISO 9000 is expected in the environmental management area.

2

The Background
to the Standard

2.1 Another BSI first

BS 7750 is another important 'first' for BSI, the British Standards
Institution, who pioneered the first quality management standard BS 5750,
which became the model for ISO 9000. It is perhaps fortunate that BSI had a
long association with environmental standards, going back over thirty
years, as many other countries will find that environmental matters come
under government departments and environmental agencies. BSI was
better equipped than most to design and produce a standard, particularly
after the success of ISO 9000.

Speaking on the matter in late 1991, Michael Heseltine MP, then
Secretary of State for the Environment, said, 'A further important
development is the work of the British Standards Institution in preparing
an environmental management standard. This initiative will provide a
detailed generic model of environmental management that any organization can use to develop its own internal management systems. It should
thus fully complement the framework that the EC regulation is expected
to provide and indeed should provide a means by which companies,

wishing to participate in the Community's Eco-audit scheme could comply with some of its requirements.

'The approach is based on the successful work of BSI in developing the BS 5750 standard for quality management. That standard was subsequently adopted as the basis for an international standard, ISO 9000. Clearly, there would be much to be gained for British business if the BSI environmental standard could again serve as an international starting point.'

Shortly after this, Bernardo Delogu of the EC Commission's XI Environment, stated, 'If a UK standard is soon established, it is clear that it could later be easily transferred into a European standard – it could even be more important than ever before, being the obvious reference for future developments at EC level.'

BS 7750 was launched on April 6 1992 under the title *BS 7750 Environmental Management Systems*, the first standard on the subject to be produced anywhere in the world. At the same time, BSI announced development plans for the standard and announced also that it would be compatible with European and international activities.

At the launch it was described as a standard which would provide a model management system for all types of organizations wishing to take a systematic and integrated approach to environmental management, with the final aim of improving their environmental performance. The press release went on to say: 'It adopts a quality system approach to environmental issues and BS 7750 is strongly linked to existing quality management practice as embodied in the series of standards known as BS 5750/EN 29000/ISO 9000.'

The companies and other bodies involved in the drawing up of the standard make interesting reading. The committees involved were the Environment and Pollution Standards Policy Committee, which, in turn, entrusted the task of drawing up the standard to Technical Committee EPC/50, upon which the following bodies were represented:

> Association of Certification Bodies
> Association of County Councils
> Association of Environmental Consultants
> BEAMA Ltd
> British Coal Corporation
> British Gas plc
> British Independent Steel Producers' Association
> British Paper and Board Industry Federation
> British Railways Board

Chemical Industries' Association
Confederation of British Industry
Consumer Policy Committee of BSI
Department of the Environment (Her Majesty's Inspectorate of Pollution)
Department of Trade and Industry
Department of Trade and Industry (Laboratory of the Government Chemist)
Engineering Employers' Federation
Environmental Council
European Resin Manufacturers' Association
Health and Safety Executive
Institute of Environmental Assessment
Institute of Wastes Management
Institution of Chemical Engineers
Institution of Environmental Health Officers
London Chief Environmental Health Officers' Association
Loss Prevention Council
Mechanical and Metal Trades Confederation
National Accreditation Council of Certification Bodies
National House-building Council
Power Generation Contractors Association
Royal Society of Chemistry
Society of Chemical Industry
Society of Motor Manufacturers and Traders Ltd
Trades Union Congress
United Kingdom Petroleum Industry Association Ltd
Water Services Association of England and Wales

2.2 The overall position of the standard

Other books explain the differences between product standards, management system standards and such elementary functions as calibration and measurement. (See, for example, my own *ISO 9000*, also published by Gower.) This is the second management standard published by BSI, and as the first, BS 5750, has already become the internationally accepted ISO 9000, all the signs are that this standard will also form the basis for an international environment standard with an ISO number, or simply become that standard. When, and if, the EC Eco-audit regulation is implemented, the comprehensive nature of this standard is such that it should completely satisfy its demands, indeed, a careful analysis of both reveals that the BSI standard goes even further than the Eco-audit requirements.

2.3   How it will work

The standard specifies the requirements for implementing and maintaining an environmental management system which can also demonstrate that it complies with company environmental policy, which, in turn, specifies compliance with relevant regulations, as a starting point.

In practice this means that a company will document the evidence that it is aware of regulations, and build a management system which can ensure compliance with those regulations, and finally produce evidence of that system for inspection.

This, of course, assumes that a company will also instal all the operating and control procedures required in such compliance, whether they be the re-cycling of waste or the scrubbing of emissions from chimney stacks. The two main assurance elements of the standard are assurance to oneself that one is in compliance with a stated environmental policy, and a demonstration of that compliance to others.

The demonstration required without the standard, or before it existed, had to be in the form of a vendor assessment carried out by demanding customers in the form of an in-house inspection. If several demanding customers insisted on demonstrations of environmental standards, this meant several in-house inspections. BS 7750 is a third party verifiable standard, which should eliminate the need for multiple vendor assessments.

In the language of modern manufacturing, everyone has a customer, so each company in a supply chain will eventually pass the demand for the standard down to its suppliers. It is fundamental to achieving the standard that it be imposed onto suppliers delivering component parts or materials into one's own assemblies or processes.

A summary of how it will work could read as follows: a company will make a policy decision to go for BS 7750, will ensure that it has a register of appropriate environmental regulations and an agreed environmental policy, will organize itself to implement and maintain an environment management system and will demonstrate compliance to the standard to an independent inspection agency.

The press release mentioned in Section 2.1 above stated that the standard 'does not specify expected levels of organizational performance but rather specifies a standardized management system that is capable of independent assessment and verification.' Accreditation of any certification scheme linked to the standard will be controlled by the Department of Trade and Industry (DTI).

The press release continued: 'With a view to European developments, the new standard is currently compatible with the European Community's proposed regulation on environmental auditing.' The regulation has now passed through the European Parliament and awaits adoption by the European Council. Any changes will be fed back into the standard in 1993. As with ISO 9000, the development of industrial sector codes, sometimes called 'specific regulations' are planned for specific industries in the application of the standard.

The publication of BS 7750 was supported by the UK Department of Trade and Industry (DTI), and the Department of the Environment (DOE). This is very interesting and contains an instructive message for other countries. The accreditation of the certification scheme for the standard was the responsibility of the Department of Industry, not Environment, and both of these government departments were involved with the standard also through the ACBE, the Advisory Committee on Business and the Environment, a special body set up jointly by the DOE and DTI in order to create closer links with business on national environmental policy.

This should be of extreme interest to other countries now intending to adopt the standard. Each national standards authority can simply deal with BSI in the normal way in 'the adoption of the standard and eventually the ISO or EN version will ensure conformity within the EC. But what about the national certification schemes? Is this an industrial or an environmental matter? We can be sure that it will become highly political and one thing is clear. If the level of competence is low in a national environmental agency or in a national environmental government department, and that agency or department is allowed to take responsibility for the national accreditation scheme, then that country will be at a commercial disadvantage as accreditation becomes a means of marketing enhancement. Some European environmental agencies have still to demonstrate their competence in this area.

The spread of ISO 9000 provides dramatic evidence of the influence of national certification schemes on the penetration of the standard. While the UK forged ahead, with more than 10,000 registrations at the time of writing, other countries lagged far behind because of foot-dragging by those responsible for national accreditation schemes, and even by government departments responsible for information about the demands of the Single Market and other major overseas markets.

In addition to this there is the possibility that members of some national environmental agencies may even be anti-industry and this will certainly lead to a highly-charged political dimension. All of these issues will be resolved within the EC by the legal status of harmonized standards in the Single Market. If your local agency cannot yet certify you to BS 7750, get BSI, or one of the private assessors it recognizes, to do it for you. And this can apply to companies in the US, Canada, Australia and the Pacific Basin. It will cost more and be a little less convenient than obtaining help and certification locally, but it is how the Single Market is going to work, and why this standard is destined to become a European and, possibly, an international norm.

2.4 Demonstrability and demonstration

Any company can claim that it is both quality controlled and environmentally caring, and how often in the past without independent third party accreditation one has heard such claims. Indeed, prior to real third party verification the words 'quality' and 'environment' were being demeaned, because of the many unverifiable and insupportable claims for both.

Accreditation arose out of stringent vendor assessment schemes carried out by large demanding buyers, where the buyer, in effect, demanded demonstration at shop floor level of the control mechanisms. The first national quality management standards like BS 5750 created real third party verification, as distinct from the first and second parties of customer and supplier. There were also other good commonsense and ethical reasons why procurement in particular should be protected by audits on the quality of the product, not the least of which was to safeguard against favouritism in choice of suppliers, which also led to abuse and fraud, as well as to loss of quality in component supply, and, in industries like aviation, to potential disaster.

Demonstrability is also an open demonstration between staff members to each other, from vendor to main manufacturer and from manufacturer to customer. For ISO 9000 it was just between these, but other parties are now also involved, such as the 'interested parties' mentioned in the environment standard, in effect the outside community, and the 'stakeholders' discussed in the latest versions of the ISO 9000 standard. The environment standard now asks for the openness to be extended to such stakeholders as the community. This is no less than a demonstration that one has nothing to hide.

Demonstrability begins at the top with the decision to adopt a policy on quality, and to set up the project team and assign the necessary resources. Its successful completion begins with the application to the national accreditation agency for certification, and is completed in the ultimate act of demonstration of the hanging of the framed certificate on the wall of the company's front foyer. The best companies now hang a framed triptych announcing quality, environment and health and safety conformance.

2.5 The EC Eco-Management and Auditing Scheme

From its inception the designers of the standard were aware of the evolving EC Eco-Management proposal and carried out the design with this in mind. Information was exchanged regularly between BSI and the Commission's DG XI to the point where the European proposals included reference to standards and certification schemes based on the British standard, including even text from BS 7750 itself. To quote BSI on the subject, 'Both documents are thus designed to work "in concert". The European proposals focus on auditing, the internal management systems, publishing the audit results, and verification. The British Standard is complementary in approach, designed to ensure that appropriate management systems to be audited are in place.'

The European Parliament has passed the Eco-Management and Auditing Scheme regulation and the European Council is expected to have adopted it by the time this book is published.

The regulation named in the final proposal for the Eco-Management scheme appears in the title thus: 'Proposal for a Council Regulation allowing voluntary participation in a Community Eco-Management scheme.' The reference to the document is COM(91) 459, published in the Official Journal of the European Communities, series C. The regulation defines the issues, criteria and requirements to be considered in the internal environmental protection system along the following lines.

European member states must establish a voluntary Eco-Management and Auditing Scheme. Companies in certain industrial activities that wish to participate must commit themselves to do the following:

- Establish an internal environmental protection system.
- Evaluate objectively the environmental performance of the system.
- Inform the public on environmental performance.
- Have the public statement verified by an accredited environmental auditor.

The EC position is that the Eco-Management scheme will be operated using European standards, the only one of which, apart from auditing standards, is BS 7750, which relates to an environmental management system and a related certification scheme. As with ISO 9000, those companies which comply will be registered and allowed to use a logo which relates to those sites which have been verified to comply under the scheme.

For readers new to quality management standards it may be worth stressing that the voluntary nature of this regulation is a legal definition only as far as both private and public buyers will begin to demand its implementation by their suppliers at an ever increasing rate.

Chapter 5 looks in more detail at the EC Eco-Management and Auditing Scheme.

2.6 The global perspective

In July 1991, within a month of the publication of the first draft of BS 7750, the ISO (International Organization for Standardization) formed a group called Strategic Action Group on the Environment, or SAGE, to develop a policy for ISO on environmental issues. One of the tasks of SAGE was to recommend a programme for ISO to present to the 1992 Earth Summit in Brazil, including a standards programme. Four elements were involved, which were environmental labelling, auditing, management and performance.

3

The Demands of the Environment

This chapter is designed to give readers an overview of the nature and scope of the environmental regulations which the standard will demand to be on record in the first instance in the register of legislative, regulatory and other policy requirements 'pertaining to the environmental aspects of activities, products and services'. This is known as the Register of Regulations.

The examples given here are from EC and national European regulations, but US, Canadian, Australian and other readers should simply apply the relevant legislation in their own countries to this chapter.

The elements covered here are:

- Air emissions.
- Water resources.
- Water supplies and sewage treatment.
- Waste.
- Nuisances.
- Noise.
- Radiation.
- Amenity, trees and wildlife.

- Urban renewal.
- Physical planning.
- Environmental impact assessment.
- Product use.
- Materials.
- Energy.
- Public safety.
- Staff health and safety.

3.1   Air emissions

There are two main areas of activity affected by regulations on emissions to the air. The first relates to plant or factory operations, while the second relates to transport fleets. Most of these regulations are legal requirements, which means that these standards are supported by statutory instruments.

Typical legislation applying to the air emissions of plants includes regulations concerning the length of time during which smoke of varying densities may be emitted, and emissions of sulphur dioxide, suspended particulates, lead and nitrogen dioxide. Regulations also apply to the use of smokeless heating systems in built-up areas, and to the sulphur content of gas oil. The EC council regulation No. 3322/88 relates to certain chlorofluorocarbons and halons which deplete the ozone layer.

3.2   Water resources

The regulations in this area are likely to contain both a general prohibition against water pollution as well as provisions for the licensing of direct and indirect discharges, in addition to water quality standards and those of managing water resources.

Water here can mean both inland waters and ocean. Specific regulations make it an offence to deposit deleterious matter, usually defined, in waters. Harbour waters and territorial waters are protected from dumping or discharges, while a number of regulations refer to the operation of ships and tankers, to oil discharges and to dumping at sea.

Certain substances also attract regulations. Examples of these are cadmium, mercury and hexachlorocyclohexane.

3.3 Water supplies and sewage treatment

The regulations in this area apply both to water supply companies and to private enterprises. They prohibit the contamination of any stream or reservoir used as a public water supply or any aqueduct or any other part of the supply system. It is an offence to allow pipes and other related devices to be out of repair and to cause or allow contaminated water or impurities to enter the water supply system.

Companies may feel that they have little responsibility here, but even the state of their own cisterns used for the supply of internal drinking water will be covered by a local authority regulation, which must be both in their Register of Regulations and under control to satisfy the standard.

Various water quality regulations, such as the European communities (Quality of Water Intended for Human Consumption) Regulations, also remind us that food and drinks manufacturers must be fully aware of both the regulations covering the state of their water and the controls exercised to ensure that the required quality of the water is maintained.

A large number of chemical substances are prohibited for discharge into waste or ground water systems. This area is known as effluent or other discharges, while licences are needed for controlled amounts of less damaging substances.

3.4 Waste

A number of regulations, including several made under the European Communities Act, apply to waste. These apply as follows.

Local authorities are made responsible for the planning, organization, authorization and supervision of waste operations in their areas and for the preparation of waste management plans. They may also issue permits to treat, store and tip waste.

The 1982 EC toxic waste regulations make local authorities responsible for the planning, organization and supervision of operations for the disposal of toxic and dangerous waste in their areas and for the authorization of the storage, treatment and depositing of such waste. Local authorities are required to prepare special waste management plans. The regulations provide for permits to store, treat or deposit toxic and dangerous waste. Local authorities are also responsible for the planning, organization and supervision of operations for the disposal of waste oils and the authorization of disposal arrangements. They can also provide permits for the disposal of waste oil.

Although companies may be involved with their local authorities in all of the above, the following regulations may apply directly to individual companies in the EC, with similar legislation elsewhere.

The first are the European Communities (Waste) Regulations 1984. These provide for the safe disposal and transformation operations necessary for regenerating polychlorinated biphenyls, polychlorinated terphenyls and mixtures containing one or both substances.

The European Communities (Transfrontier Shipment of Hazardous Waste) Regulations 1988 provide for an efficient and coherent system of supervision and control of the transfrontier shipment of hazardous waste. These regulations apply both to exports and imports. They prohibit holders of waste from commencing a transfrontier shipment until a notification has been sent to the appropriate authorities and duly acknowledged. A consignee in the state is prohibited from accepting hazardous waste from outside the state for disposal in the state unless it is accompanied by the appropriate acknowledgment. Carriers are prohibited from handling such wastes unless the waste is accompanied by the appropriate documentation.

Chemicals and pharmaceutical companies amongst others are affected by various national and international Dumping at Sea Acts. These in turn refer back to the Convention for the Prevention of Marine Pollution by Dumping from Ships and Aircraft, 1972 (the Oslo Convention) and the Convention for the Prevention of Marine Pollution by Dumping of Wastes and Other Matters, 1972 (the London Dumping Convention).

The acts prohibit the dumping of substances and materials from all vessels, aircraft and marine structures, for example, offshore platforms, anywhere at sea unless such dumping is carried out under and in accordance with a permit issued by the appropriate local authority.

Dumping at sea outside of territorial waters is permitted only if such dumping takes place under and in accordance with a permit granted by another state that is a party to the London or Oslo Conventions.

Other national regulations in the waste area cover local litter regulations, certain kinds of advertising, and the disposal of abandoned vehicles.

3.5   Nuisances

There are a number of national regulations under this heading which are relevant to companies in both manufacturing and service sectors. They deal with a wide variety of nuisances affecting public health and the general environment. Typical nuisances could be offending pools, ditches, watercourses, drains, accumulations and deposits. Temporary dwellings, such as caravans, lean-tos and vehicles may be included also.

3.6   Noise

Regulations in this area are also of great relevance to both manufacturing and service companies. Firstly some EC examples.

European Communities (Construction Plant and Equipment) (Permissible Noise Levels) Regulations 1988 give legal effect to EC directives on the approximation of the laws of the EC member states relating to the permissible noise levels of construction plant and equipment, for example, compressors, tower cranes, welding generators, power generators, powered hand-held concrete breakers and picks, hydraulic excavators, rope operated excavators, dozers, loaders and excavator-loaders, designed for use in or about civil engineering or building sites.

European Communities (Lawnmowers) (Permissible Noise levels) Regulations, 1989 give legal effect to EC directives on the approximation of the laws of the member states relating to permissible noise levels of motorized mowers.

3.7   Radiation

There are a number of EC and national regulations covering the important matter of control of radiation.

Council Regulation (Euratom) No. 3954/87 of 22 December 1987 lays down maximum permitted levels of radioactive contamination of food-stuffs and of feeding stuffs following a nuclear accident or any other case of radiological emergency.

Council Regulation (EEC) No. 3955/87 of 22 December 1987 sets out the conditions governing imports of agricultural products originating in third countries following the accident at the Chernobyl nuclear power station. This expired in 1989 but is an example of a regulation which could be repeated in the case of a nuclear accident.

European Communities (Medical Ionizing Radiation) Regulations, 1988 provide that all those engaged in the use of ionizing radiation for medical and dental purposes must be competent in radiation protection measures and have appropriate training. The exposure of a patient to ionizing radiation must be medically justified and the dose to the patient must be as low as is reasonably achievable.

3.8 Amenities, trees and wildlife

There will be a large number of national regulations covering trees, parks, amenities, landscape and wildlife. The following examples of the concerns of those regulations show how companies may be affected.

Companies may find that they are obliged to trim and cut hedges and trees in some instances and are prohibited to do so in others. Certain categories of trees are exempted from any cutting. In making plans for development, companies have to be aware of regulations concerning the preservation of specified trees and other amenities, including spaces, and of requirements to plant trees, shrubs and other plants and for the landscaping of structures or land.

There are numerous wildlife regulations, the objectives of which are the protection and conservation of wild fauna and flora and for the conservation of areas having specific wildlife values. Apart from the regulations which demand protection of local wildlife, others may face companies which are involved in any way with materials or components made from the fur, skins or parts of wild animals, birds or certain species of flora. Even common lizards and newts are involved.

3.9 Urban renewal/site dereliction

A company may find itself in two different positions. Firstly where it controls buildings or lands designated for urban renewal, or, secondly where it has derelict sites or buildings in dilapidated or ruinous condition and which may injure the health or safety of the neighbourhood.

A company may also own or occupy designated buildings listed for preservation for historical, cultural or architectural reasons.

3.10   Physical planning

All companies engaged in acquiring, constructing and expanding buildings must operate under physical planning regulations managed by local authorities. These lay down the requirements for permission, plans, sizes, heights and so on.

Within these are the important Environmental Impact Assessment (EIA) directives. These are fundamental to the subject of this book as they relate to the planning permissions of any development above a certain cost threshold.

The EC EIA Directive (85/337/EEC) applies to certain public and private projects and requires that those which are likely to have significant effects on the environment be subjected to assessment of such effects before development consent is given for them.

3.11   Environmental impact assessment

If one undergoes an environmental impact assessment study and fails to obtain planning permission, then the subject of operating to BS 7750 becomes irrelevant. If, on the other hand, one is facing an assessment for a new plant, this is an excellent time to begin planning for the standard as the work for the assessment can form the basis for both the proper future operation of the plant and its monitoring.

The local authority responsible for issuing planning permission for the project usually commissions an independent third party to carry out an environmental impact audit on the likely effects of the proposed project. The auditing body may use both baseline studies and predictive techniques to calculate the effects of the proposed project on such elements as air and water, quality, health, nuisance and noise.

Council Directive (85/337 EEC) sets out the projects subject to assessment. They include virtually all plants in process industries, those in extractive industries such as mining, energy producing installations and metal processing. Also included are chemical, food, textiles, wood, paper, leather, rubber and certain infrastructural projects.

The likely outputs agreed between the project proposer and the assessment agency can, if they fall within the limits set by regulations, become the targets for achievement within the subsequent operating environment. These can be set as limits beyond which the company does not intend to stray, while the limits set by the regulations can be the extreme upper limits. Staying within these and preferably at the level set in the assessment, or even lower, can be the levels set for compliance with the standard.

3.12 Product use

Product use is a most interesting issue: the reader shall be aware that this interpretation derives from an early reading of the standard, shortly after its publication at a time when companies were only just considering this aspect of the standard. However, even if the authors of the standard did not specifically have product use in mind, it is probably the most important issue for many companies, as the safety, ergonomics and 'usability' of products are certainly now legal issues, covered increasingly by legislation. This matter is made clearer in other sections of this book dealing with health and safety regulations and the use of software in on-line processing situations where there is a potential for accidents. Any reading of BS 7750, or indeed the latest version of ISO 9000, should offer convincing proof that these standards are a way of managing the end-use of products and services, that is ensuring that they are used safely and properly, or at least educating users in these respects, as far as possible. It should also offer the assurance that as a method of demonstration of a caring management system, one, at least, obtains the protection of a strong defence in cases of claims under product liability or charges of negligence.

3.13 Materials

Both the preparatory review and the subsequent controls should include use of materials as an issue. The review of materials needs to begin at the design of the product or service and its packaging, and ensure the careful use of materials, with the emphasis always on a reduction in the amount of materials used and the possibility of re-cycling. In Chapter 17 we will see a good example of how Hewlett-Packard instructs its customers in the re-cycling of a used product. Such instructions can apply both to straightforward reduction of materials, such as glass and paper re-cycling, and to the re-cycling of dangerous wastes, such as batteries.

3.14   Energy

It is essential to design and implement an energy conservation programme, both for normal processing and for all other activities, including office and transport, within the company. This area is so well developed that if one does not already have such a system, plenty are available off the shelf, or from the nearest national energy saving information agency. The standard demands that energy be treated as an issue and controlled accordingly. The preparatory review should encompass energy use and the controls should be a part of the Environment Management System, as are all the other issues.

3.15   Public safety

If there is a public safety issue, as there will be in many process plants, including chemicals and power stations, and in organizations distributing dangerous substances, a separate hazard and emergency control system, with all the accompanying emergency procedures, must be put in place. This is of such fundamental importance to both public safety and corporate and personal responsibility that unless the industry has codes of practice, control mechanisms and its own independent or government audits, an expert outside agency should be used. This issue makes an interesting point about the standard. The standard is a control and documentation mechanism, which at all times assumes expert management and control at every level. It expects aircraft to be flown by pilots and to be maintained by engineers, who are, in turn, monitored by a management system.

3.16 Health and safety

Health and safety are very much a sub-set of the environment management system, but they are now compulsory in many parts of the world. Because of this, and because of the ease with which they can be fitted into, and accommodated by, the environment management system, they are, in fact, an excellent motivation for adopting BS 7750.

They need be treated no differently from any other issue and can be controlled accordingly. They are discussed in detail in the next chapter.

3.17 General note on the international position

In the United States and Canada, companies respond to a combination of regulatory and voluntary measures involving these issues. In some cases shareholders demand compliance with environmental standards as, increasingly, investment seeks 'green' activities.

The main recommended practices appear to be:

- A 'Compliance' auditing scheme operated by the EPA (Environmental Protection Agency).
- A chemicals industry 'Responsible Care Programme'.
- A 'Toxic Release Inventory Act'.

The EPA scheme is still voluntary (at the time of writing) but, as in Europe, its adoption may be mandatory in order to obtain local public authority licences or satisfy local regulations. Some quite comprehensive protocols are published by the EPA in both the *Environmental Audit Program – Design Guidelines for Federal Agencies*, EPA 130/4–89/001 and *Generic Protocol for Environmental Audits at Federal Facilities*, EPA/130/4–89/002.

There is also a legal responsibility to report the emission of certain toxic substances into air or water, or in discharges to water or land. This affects around three hundred substances used in manufacturing or in applications such as cleaning, and around 20,000 companies appear to have been complying. This act has caused companies to clean up their operations, and now at least twelve states are using the regulation to reduce emissions through legislation. Other new laws and clean air acts continue to emerge.

The 'Responsible Care Programme' may have influenced both BS 7750 and the EC Eco-Management regulation.

In Sweden a 1991 law demands that at least 6000 industrial enterprises make annual returns on their performance in complying with existing environmental regulations. These reports are made public. These measures are also supported by voluntary industry inspections and audits.

In France self-inspection schemes are employed in many establishments, while the Ministry of the Environment has been publishing the names and locations of polluting companies. The Netherlands have persuaded industry to introduce an 'Internal Environmental Protection System', which reads like the BS standard. In both the Netherlands and France mandatory controls seem likely.

There are still no legal requirements in Europe to introduce a formal environment management system. This will change if the four year voluntary stage of the Eco-Management scheme, discussed in Chapter 5 moves to a mandatory stage, but it is already changing in practice as health and safety regulations, which need an overall system of management, become law, and as individual state and local authority environmental regulations also demand demonstrable control.

4

The Health and Safety Regulations

As the environmental standard emerged, health and safety regulations were being implemented by national government legislation around the world. Within the European Community a large number of directives emerged to be implemented by national Acts of parliament.

What follows is based on EC directives at the time of writing, but the reader is reminded, as elsewhere, that the examples of specific legislation and regulations in this book are merely intended as a guide. There can be no substitute under the environment standard for building one's own register of regulations. This is of particular importance to readers in the US, Canada, Australia and other non-European territories. The EC samples are a good guide, but both EC and non-EC readers must establish their own regulations; even EC readers may find that the regulations in this book have been amended by the time it is read.

4.1   Environment and health and safety

The standard defines 'environment' as including human and other 'living systems' within the surroundings and conditions in which they operate. It also asks that the environmental policy be consistent with the occupational health and safety policy. These two statements are enough in themselves to warrant the inclusion of health and safety under environment, but there are several other good reasons, one being that the workplace is indeed part of the environment and as such can be the same as, better than, or worse than the outside environment, depending upon the measures taken, for example, protection of staff, height of chimneys, ventilation, noise.

A second reason is that while much of the environmental demand is still voluntary, all of the relevant health and safety regulations are now mandatory and enacted in legislation. Last, but not least, the health and safety procedures need controls, demanded by legislation: the environment standard and its control system is the ideal vehicle for the incorporation of the health and safety demands and for the satisfaction of health and safety inspectors.

The close relationship between health and safety and environment at times causes problems in distinguishing between the two. Packaging legislation comes into both areas, as do dangerous chemicals. A computer or software manufacturer will find VDUs are a health issue with internal staff, but an environment issue with customers. The only reason for keeping them separate is that personnel or human resources management will have a direct responsibility for implementing and controlling the health and safety procedures; the quality manager will still control the overall environment and quality management system, which in turn controls health and safety.

4.2   The scope of the regulations

There are new health and safety regulations on the way (see Section 4.3) and there are already a large number of regulations in existence. One overall EC 'Framework' directive set the scene for most of the new and forthcoming specific regulations. This was the 1989 Framework directive on health and safety which became a Safety, Health and Welfare at Work act in member states. This existing piece of legislation still stands, but is modified or given more substance by the specific new directives.

Even before the 1989 legislation there was a large amount of existing legislation. In the EC, for example, there are thirty-two regulations based on pre-Single European Act EC directives with health and safety implications. Twenty of these relate to the classification, packaging, labelling, marketing and use of dangerous substances and preparations. Some of the new directives update these. Three regulations fall under the infamous 'Seveso Directives' and relate to emergency planning at chemical plants. These arose from the Italian Seveso disaster. A number relate to the certification of electrical equipment while one, about to be amended, deals with safety signs in the workplace. Four concern chemical, physical and biological agents at work.

Returning to the Framework directive and its member state equivalent Safety, Health and Welfare at Work acts, we find very broad and general demands protecting 'so far as is reasonably practicable' the safety and health at work of all employees. Employers are asked to provide information, instruction, training and supervision 'as is necessary to ensure so far as is reasonably practicable the safety and health at work of their employees.' This rather generous scope is not, however, repeated in the specific regulations, which are listed in the next section.

Employees are also asked to take responsibility for their health and safety, and manufacturers, designers and importers 'of any article or substance' must ensure that they are without risk to health and safety and that users are supplied with adequate health and safety information. Places of work should be safe and without risk to health.

Employers must prepare a Safety Statement which specifies the manner in which the health and safety of employees shall be secured at work. This has caused much confusion and it is hoped that a standard such as BS 7750 will remove the need for this rather vague demand. Readers should take care, however, to go on producing a Safety Statement for as long as it is requested, as it may take time for state health and safety inspectors to realize the implications and benefits to them of the standard.

The overall act also demands that employees have the right to appoint a Safety Representative from amongst their number. This appears to fly right in the face of the Environmental Management Standard and ISO 9000, both of which ask for a clearly defined 'management representative', such as a quality manager, to be named as responsible for the overall task of coordinating the complex management system which manages quality and safety. This, however, does not exclude a staff representative, but does clearly allocate responsibility. As we are dealing here with potential liability, and even the possibility of criminal charges in the case of neglect, this matter will not be delegated by management.

4.3 The detailed directives

These directives are mainly post-Single European Act regulations. All of these have or will result in specific legislation in the individual member states.

> Workplace Directive
> Work Equipment Directive
> Personal Protective Equipment Directive
> Handling of Loads Directive
> VDU Directive
> Exposure to Carcinogens Directive
> Provision of Safety and/or Health Signs at Work
> Safety and Health Requirements at Temporary or Mobile Sites
> Safety and Health at Work of Pregnant Workers and Workers who have
>   recently Given Birth or are Breastfeeding

At the time of writing, directives were expected for working hours, young people at work, mines and quarries, offshore, vessels, transport of dangerous goods, exposure to dangerous substances, activities in the transport sector, fairgrounds and playgrounds.

4.4 The safe workplace

If the workplace is a new building to be used for the first time after 31 December 1992, twenty requirements must be satisfied for health and safety. In most countries local building regulations will have already made these law. Looking down the list for possible new requirements one sees the need to provide facilities for pregnant and breastfeeding women, changing rooms, and facilities for the disabled. When we look at the requirements for existing buildings, however, we find that all these too must be added to the buildings. The actual new requirements for new buildings after 31 December 1992 relate to floors, windows, escalators, loading bays and overcrowding.

Once the structure is in place and satisfies the directive, the maintenance of equipment is called for, together with ventilation, temperature, lighting, traffic routes, toilets, and other relevant issues.

Work equipment must be safe and meet guidelines laid down in the work equipment directive. This involves training also. Users of VDUs are protected by a number of measures – keyboard, design, radiation, task and software design, timing, eye protection, and training.

Workers lifting loads will do so properly and only of the weights allowed by their abilities. They will be trained in lifting. Personal protective equipment will be available and be in use wherever required and there will be full training in its use.

Safety and health signs will be displayed where and as they are required, and there will be first aid facilities, and both facilities and procedures for pregnant and breastfeeding women. All dangerous substances will be handled according to strict regulations.

4.5   The Safety Statement

Both the individual country acts and the EC directive call for a Safety Statement, which has given employers some problems. In a number of cases insurance companies are not happy with a statement which they privately believe is not worth the paper it is written on. The problem is that most Safety Statements are indeed not worth the paper they are written on, if they are not related to a system of management control.

The Safety Statement should be a copy of the Quality Policy Statement, relating to management's intention to operate a control system to ensure that all health and safety regulations, and all other health and safety targets set by policy, will be implemented, monitored and controlled, just like the quality statement, but as we have already noted, until health and safety inspectors realize this, one may have to go on supplying them with this legally-demanded device. It should be filed in the Register of Regulations.

5

The EC Eco-Management and Auditing Scheme

Generally, the distinction between a directive or regulation and a standard is straightforward. The standard allows a directive or regulation to be met, and a management system developed to a standard such as ISO 9000 or BS 7750 may allow whole sets of regulations and directives to be satisfied.

The Eco-Management scheme (the full title of this regulation is 'The Eco-Management and Auditing Scheme) is somewhat more complex. Firstly, although it is voluntary, for four years at least, for companies which aspire to it, it is a 'regulation' which, within the EC sense, has bound member states to set up national schemes to support it from 1 January 1993, after which date the regulation came into force. Within twelve months from that date the certification schemes were to be in operation and the regulation 'applies with effect' from 1 July 1994.

It is even more complex, however. It is so detailed in its demands that it specifies almost as much as a full standard, not quite enough, however, when one looks closely. Many companies in Europe may simply take the Eco-Management scheme as both a voluntary directive and a standard and implement it. Indeed, this whole book could be re-titled *Eco-Management and Auditing Scheme* and the contents used as instructions for its implementation.

Having looked hard at both the Eco-Management scheme and BS 7750, however, and particularly at the clear definition of key elements in the latter, this writer has no hesitation in recommending BS 7750 as the ideal way of meeting not only the British Standard, which is also the basis for the international standard, but for satisfying the demands of the Eco-Management scheme as well.

5.1 General outline of the Eco-Management scheme

The central demand of the Eco-Management scheme is for an 'environmental protection system'. This is the counterpart of the BS 7750 'Environment Management System'. This applies to a specified production site and involves a 'systematic, periodic evaluation of environmental performance'.

Everything else demanded by the scheme, from a formal stated policy to a programme of site measures and a management system is more than catered for in BS 7750. One area of different emphasis, described as a 'key element', is an environmental audit. Audits are called for in the BS standard but the Eco-Management scheme asks that this key audit element be followed up with an 'environmental statement'.

This is not as one might expect a mere policy statement, but a full document describing the company's performances and intentions, 'formally validated by an *accredited independent auditor*' (see page 116), on the basis, inter alia, of the internal audit conducted and its results. And it goes further, for this so-called statement must be made available to the public.

5.2 The motivation and methods

In the words of the Eco-Management scheme the system is designed 'to target motivated companies straightaway', to improve their environmental performances.

Another statement is significant: 'The voluntary nature of the proposed approach is perfectly consistent with the objectives of the Commission which wishes at this stage to give an impetus to improvements in the performance of industrial operations, irrespective of size or nature, to encourage companies to behave responsibly and to facilitate the flow of information to, and participation by, the public, as an addition to the traditional regulatory approach.'

One can gather from this that the Eco-Management scheme, like BS 7750, needs to start somewhere and that a certain element of proselytizing will prevail at this 'missionary stage'. All the signs are however, that, as it takes hold and spreads, the voluntary nature will give way to either mandatory market pressures or regulations.

It goes on to say that the 'credibility of the voluntary scheme is guaranteed by systematic external verification carried out by an independent accredited auditor'. EN 29000, the European norm for ISO 9000, is suggested as the standard's structure. At the time of writing it could only be presumed that ISO 9000 will be expanded to embrace BS 7750 or its ISO or EN equivalents when they evolve.

5.3 Principal features

The Eco-Management scheme is open to 'any company engaged in industrial activities at a given production site'. This can be taken to apply also to service and distribution companies.

The companies which participate, and the word 'participate' is used throughout the regulation, as if participation also embraces certification, will be granted a logo for the 'specific site' participating. This is followed by a confusing clause which reads, 'which may be used on condition that the exact meaning of the logo, which proves nothing about the product's intrinsic quality, is readily understood'. This raises more questions than it answers. One can easily understand that the logo must not be used on the product – it refers to the environment system at the site, and the product might not even be desirable. Requesting that the exact meaning of the logo be understood may appear superfluous to a company which has just introduced a rigorous environment management system.

The participating companies will be published in the official journal of the EC, a kind of honours list, or EC-wide register of members.

The confusion caused by the strange clause about understanding the logo may be put to one side at the point where the Eco-Management scheme specifies the four conditions for 'participation'. All participating companies must satisfy these conditions:

- Carry out an environmental review of the site concerned.
- On the basis of the findings, introduce an environmental protection system, including the elements already mentioned and an audit procedure.
- After each audit, draw up an environmental statement intended for the public, the contents to include figures on performance, problems brought to light, policy objectives and future intentions.
- Have an accredited environmental auditor examine and validate the audit and the statement.

This is what each participating company must do, but from 1 January 1993 each member state was obliged to put the following paragraph into effect:

Each member state will ensure that steps are taken to introduce an accreditation system, responsible for establishing and applying appropriate procedures for the accreditation of the environmental auditors entrusted with the task of validating the audit process and the environmental report. The member states may rely on existing bodies, but must observe the conditions and criteria specified by this proposal in the matter of competence and pluralism.

5.4 Difficulties with the Eco-Management scheme

At first glance, this fifty page document appears comprehensive, covering all the elements one would expect in a regulation laying down guidelines for meeting its objectives. On reflection, however, if one tried to implement a system based on the information contained within this regulation only, one might experience extreme difficulty.

For one thing, while it emphasizes environmental measures, it makes virtually no reference to how one finds out what to do, whereas BS 7750 deals explicitly with a Register of Regulations. With the British standard one begins by understanding the rules and demands of the environment and the community, and then, one sets out to satisfy them.

The Eco-Management scheme appears to concentrate so heavily on the concept of 'audit' that it neglects identification of the issues to be audited, whereas the British standard makes sensible statements about moving from the realization and identification of regulations in the general sense to the specific issues. A very good example of this is that while air and water regulations may have widespread implications for industry, individual companies will be involved only in specific 'issues', so issue identification is an important activity in the initial reviews.

5.5   Strengths of the Eco-Management scheme

The great strength of the Eco-Management scheme is that it is a Community-wide regulation, all of the powers of which will be in full effect without exception in the member states from 1 July 1994. As such, it is a far-seeing and revolutionary step, which will probably lead to global acceptance.

It is also an attractive concept, its logo an exciting target for industry. It is hoped it will help to eliminate spurious claims of products being clean and green and even lead to the proscription of such claims, or to their being branded as false advertising.

Most important of all perhaps is that it is forcing member states to set up environmental regulatory authorities and systems of support and accreditation for participating companies. These volunteering companies will, in fact, spearhead a movement which should become irreversible.

5.6   Current and future developments

At the time of writing, the signs are that BS 7750, either in its current form or as future ISO or EN standard, will be the single, European-wide, and even international, standard for meeting the requirements of the Eco-Management scheme. Indeed, one could go further and say that everything the Eco-Management scheme demands, and more, is to be found in the British standard. In future editions of this book we will be able to report on such developments, such as the EC requesting CEN to adopt an ISO version of BS 7750 and introducing an equivalent EN 29000-type norm.

PART II

Definition and Scope

6

Definition of BS 7750

The Foreword to the first published draft of BS 7750 states: 'The standard is designed to enable any organization to establish an effective management system, as a foundation for both sound environmental performance and participation in "environment auditing" schemes.' It goes on to state that the standard shares common management system principles with ISO 9000 and that organizations can choose to use an existing ISO 9000 system as a basis for the environment standard. Also the existing certification or registration schemes for ISO 9000 can be used to demonstrate compliance with the environmental standard. At this stage the Foreword also adds that suitable sector application guides will be needed for specific industries. Some of these are in production, more are expected, but the Foreword mentions industries with the following characteristics as needing sector application guides:

- Where there are complex environmental effects.
- A large number of constituent companies.
- Widely differing, but loosely related, operations and disciplines.
- Temporary and/on off-site activities.
- Substantial uses of sub-contracting.

6.1 The sector application guides

In the case of ISO 9000 there were also sector application guides, known as specific regulations, relating to industries or processes which had specific quality control needs, for example, process industries and software. The control of quality, however, seems to be possible with a number of generic or common rules, such as inspection of key procedures. To control the quality of a product though has fewer possible requirements than the control of the environment in the production of that product. One can produce a 'perfect' product to a quality standard and do it in a way that damages the environment, through disposal of waste for example.

This suggests a greater number of specific sectoral application guides and these should emerge industry sector by sector with agreed processes, controls, measurements and specifications. Nothing at this stage, however, suggests that an individual company cannot design its own controls as long as they relate to the conformance of activities to existing relevant environment directives.

If, however, readers are working in industries with the characteristics described above, it would be wise to refer to industry associations for information, or to share experiences with others. The standards and certification agencies will look with respect on agreed sectoral criteria. Indeed, to a great extent industry will write its own specifications for the standard and develop its own controls. However, the assessors will require proof that these help individual companies to conform with existing directives.

6.2 The scope of the standard

Paragraph 1 of the Document describing BS 7750 states that the standard 'specifies requirements for the development, implementation and maintenance of environmental management systems aimed at ensuring compliance with stated environmental policy and objectives. The standard does not itself lay down specific environmental performance criteria.' This is most interesting for those involved with ISO 9000, as it appears to mirror a fundamental characteristic of that standard, which allows enterprises to set their own criteria. It says, here is a framework, now you write in the requirements and design your controls.

However, there is a fundamental difference between the two standards. A good way to clarify this is to remember that you can specify rubbish with ISO 9000, if you are making rubbish. Imagine two standards for two different kinds of glass. One must withstand the impact of a grown man in protective clothing and helmet leaping against it, and this is for use in high rise buildings. A second standard is for glass used on movie sets, which must shatter on impact, harmlessly, if a child rushes through it. The 'quality' of the second sheet of glass or its fitness for its purpose is no less than that of the first, and each can achieve ISO 9000 if their management systems ensure continuous compliance with their respective standards.

The environmental standard, however, is not so liberal. Existing environment directives will not allow for polluted air or water in certain cases. In this respect, although at first glance we are allowed write our own specifications, we have far less freedom with our outputs than we have with ISO 9000.

6.3   Definitions

Under this heading, the standard defines all of the expressions used. This is similar to the approach taken in BS 5750 or ISO 9000 where there is a similar set of definitions used in the sphere of quality.

The environment is the surroundings and conditions in which an organization operates and it includes the 'living systems' both human and other therein. This is very clear and it includes both in-plant and external operations, such as transport. It embraces living things within the scope of the enterprise, whether they be newts in ponds on site or humans. It can be presumed therefore to include all safety and health regulations in addition to environmental regulations, which greatly increases the size of the Register of Regulations.

This is not all. The definition goes on to say that 'as the environmental effects of the organization may reach all parts of the world, the environment in this context extends from within the workplace to the global system.' This speaks for itself and illustrates the dramatic and far reaching nature of this standard, which will have profound effects when implemented. If anyone is still in any doubt about its meaning, the paragraph defining 'environmental effects' leaves no such doubt when it says, 'Any direct or indirect impingements of the activities, products and services of the organization upon the environment, whether adverse or beneficial.'

This encompasses operations, such as processing, the use of, and ultimate disposal of, products, packaging, services, transport, noise and nuisance. These in turn are now also passed onto suppliers in the way that quality demands are passed on in ISO 9000. We will look at this in greater depth later, but this has very serious implications for industries in areas such as energy and transport supply, who if their own products are not produced to the standard could be the cause of the customers who use their products not being able to implement the standard themselves.

Taken to the ultimate, this could result in a situation where major public procurement buyers in the EC Single Market, for example, could demand BS 7750 or its ISO equivalent and potential suppliers might not be able to qualify because their local electricity utility was not applying an EC directive on emissions. Here we could find a local utility, possibly state-owned, acting as a barrier to national industrial output.

A final important paragraph in the definitions section deals with 'interested parties'. This leaves absolutely nothing to the imagination and includes those with an interest in the environmental effects of an organization's activities, products and services. These include those exercising statutory environmental control over the organization, local residents, the organization's investors and insurers, the workforce, customers and consumers, environmental interest groups and the general public.

6.4 System requirements

Under this the standard lays out the structure of what is required within each enterprise for its implementation.

6.4.1 The environmental management system

The order in which this and the related paragraphs are listed may not be the best. It could be more convenient and meaningful to put policy first, organization second and this paragraph on the environmental management system next. Later when we deal with implementing the standard, it will be seen that it is recommended that commitment and policy are sought first, as that is how it actually works in practice. To adhere to the standard, however, and to make the reading of it more meaningful, we will stay with its order for the purposes of this chapter.

The environmental management system is the means of ensuring that the enterprise in all of its operations and activities conforms to environmental targets set by its policy, which itself relates to standards set by directives or other criteria. The system, in turn, consists of procedures, instructions and controls designed in accordance with the requirements of the standard, all of which will be covered in detail in the following chapters.

If there are industrial codes of practice already in existence, the organization must also take these into account.

6.4.2 Environmental policy

The standard asks for a defined and relevant policy and it wants it understood, implemented and maintained, and even published. It also wants it to include a commitment to continual improvement of environmental performance, which may not be very reasonable, as adherence to or doing better than the demands of relevant directives would seem to be sufficient, and good environmental practices must have practical limits. The top ISO 9000 achievers soon found that they needed something beyond the operation of a quality management system to achieve even better performances. The environment standard also will have its limitations, but not damaging the environment must in itself be a most desirable goal.

The policy must obviously embrace the stated environmental objectives as laid down both in the directives and in the management system.

6.4.3 Organization and personnel

Once the commitment and policy are in place the enterprise is expected to set up the appropriate organization and to document the relevant procedures. This is not necessarily a separate staff, although it will be required that one person, who could be the quality manager, or an equivalent environmental manager, be named for ensuring overall compliance with the standard.

Many companies who have achieved ISO 9000 are anxious to find new quality targets, and BS 7750 could provide just such a target and satisfy the quality manager in this respect. In such cases, or unless the environmental demands are so great that a separate function from operational quality is required, the quality manager seems the ideal choice for being given the responsibility for the environment standard. Apart from any other reason, he or she is ideally qualified to get the standard implemented.

Management is expected to define this and related roles and to provide the necessary resources for the implementation of the system and to put in place the control and verification procedures. Management is also expected to 'act in emergency situations', a statement that appears to be unnecessary, as implicit in the standard is the control of situations which could lead to an emergency. What it probably means is that any industry with a potential for danger to the public or to the environment must have emergency procedures involving management. This would be very relevant with power utilities and chemical companies.

There is also a strong paragraph in the standard under this section on making all employees aware of what compliance with the environmental policy and objectives means, both in terms of the results to be achieved and the part that each employee must play. The paragraph also spells out the need for training and for specifying individual roles in the management system.

6.4.4 Environmental effects

It is in this section that the Register of Regulations appears, called the 'Register of legislative, regulatory and other policy requirements'. Throughout this book it will be referred to as the 'Register of Regulations' for convenience. In it management must record all legislative and regulatory requirements, for example, current directives and government regulations concerning the environment. Chapter 3 has already listed the areas in which these will apply. The register could also contain customer environmental demands and objectives set by policy.

Under this section the company is also expected to set up and maintain procedures for dealing with 'relevant' interested parties (both inside and outside) concerning the effects of its operations on the environment. For the first time this gives the general public a right to know, but some companies may sigh with relief at the word 'relevant' here, as this may provide the means for ignoring cranks.

6.4.5 Environmental effects evaluation and register

This is a second register for the company's record of how it deals with the activities covered by the regulations. There are two parts to this. Firstly, the system which is described elsewhere as the environmental management system will establish and maintain the procedures for examining and assessing the progress of meeting the demands of the regulations or of company policy targets; secondly, those procedures 'identified as significant' will be recorded in a register.

The standard has not yet told us what goes into the register, so we may assume that it is a record of results of how good or bad we are in meeting the objectives, by significant activity. It does, however, list the possible activities to be controlled in the procedures of the environmental management system, so we may assume that these are significant enough to require inclusion in the register. They are:

- Controlled and uncontrolled emissions to the atmosphere.
- Controlled and uncontrolled discharges to water.
- Solid and other wastes.
- Contamination of land.
- Use of land, water, fuels and energy, and other natural resources.
- Noise, odour, dust, vibration and visual impact.
- Effects on specific parts of the environment and ecosystems.

The requirement goes beyond normal operating conditions and demands procedures for abnormal conditions, incidents, accidents and emergencies and even 'past activities'. The last note suggests that past dumping may not escape the standard.

6.5 Environmental objectives

A careful reading of the text in this section of the standard suggests the following. In the first instance compliance with all existing legislative and regulatory requirements is essential. This simply means that the regulations already in existence and collected in the ring binder called the Register of Regulations must be adhered to as the basic requirements of the standard. Over and above these, there can be other objectives and targets set by the company or by its customers.

In this respect the standard is very similar to ISO 9000 and this paragraph may be dangerous for the over-enthusiastic, for one can over-specify. It appears to mean that, although the minimum to be expected is the meeting of the regulations, you can go beyond these with your own targets, but should you do so your achieving of the standard will depend upon you meeting *your own* specifications. Should you specify too high, you may fail to achieve the standard; others who may specify only the legal levels may achieve it.

To balance this, the standard also asks for a commitment to continual improvement, so perhaps the sensible target is the minimum legal requirement in the first instance with a policy for improving beyond that point after the minimum has been reached.

Another cautionary note in this paragraph is that the targets beyond the required baseline should be identified 'after consideration of the environmental effects register and the financial, operational and business requirements of the organization, in conjunction with the views of interested parties'. This seems to mean that after the minimum legal requirements we ask what else is messy and can we afford to clean it up, and what else can we afford that will please our interested public?

6.6   The environmental management programme

The contents of this section of the standard may cause confusion. At first glance it appears to duplicate the environmental management system, and may well indeed do so, but upon further scrutiny it appears to suggest that the initial implementation of the environmental management system is regarded as a 'programme' and that all new projects, of which there may be more than one at any time, need to be subject to a programme. In a footnote it introduces the concept of 'environmental assessment'. This is covered in Chapter 3, Section 3.11. As for the programme for the initial introduction of the standard, or for new developments, this is covered from Chapter 10 onwards and, in particular, in Chapter 12, Making the Plans.

The programmes or 'plans' are expected to contain targets and responsibilities and the mechanisms for achieving the targets.

6.7   The Environment Manual

This is called the Environmental Management Manual in the standard, but as this is rather a formal title for what will be another popular version of the now famous Quality Manual, it is referred to as the Environment Manual in this book. This is another point at which to pause and reflect that, in addition to the documented system, there are now no less than three corporate-type manuals. The first is the Register of Regulations, officially titled 'Register of legislative, regulatory and other policy requirements'; the second is the Environmental Effects Register; and the third is the Environment Manual. All the procedures in the environment management system could also be bound in manuals and called 'Procedural Manuals', but these three are overviews or corporate in nature.

The Environment Manual will be very familiar to ISO 9000 users. It is the master manual, containing the elements of company policy, organization and management which implement and control the system which ensures compliance with the standard. Little will be said about it here as it is important enough to warrant a generic sample given in Appendix 4. The standard asks that the manual contain policy, targets, programmes, roles, system elements and documentation. It also asks for control of documentation, including updates and revisions.

6.8   Operational controls

It seems strange that the authors of the standard do not simply say that the operational control is that part of the environmental management system which controls it. Instead, operational control is presented as though it were additional to the management system. Having said this, it is perfectly clear that each company is expected to identify the activities which affect and 'have the potential to affect' the environment.

This is where we find another difference with ISO 9000. Quality is a matter to be decided between customer and manufacturer. If for some reason the customer wants a lower grade of product and specifies it accordingly, that grade becomes the required 'quality'. One does not have such discretion with the environment, nor does one have the number of potential variables which could arise in the fitness of purpose or quality of a component. There will also be a similarity in how companies treat the environment, but, most of all, regulations will set baseline limits, and, with the limits fixed, the control will become standard within industry.

The standard demands that the activities be carried out under controlled conditions. It asks that particular attention be paid to documented work instructions, which are referred to as procedures or procedural manuals throughout this book. It specifies that these cover both the company's own activities and 'by others acting on its behalf'. This raises the matter of sub-contractors and suppliers. It is clear from this that sub-contractors, both in a project and in a manufacturing situation, are the responsibility of the standard-seeker, as are suppliers. This is spelt out further in a demand for procedures dealing with procurement and contracted activities, to ensure that suppliers and contractors comply with requirements. The subject of procurement will be dealt with in more depth later.

Three more items of control are mentioned as examples:

- The monitoring and control of relevant process characteristics such as effluent and waste.
- Approval of planned process and equipment.
- Criteria for performance, to be stipulated in writing.

These are three good examples which can be summarized in general – as the monitoring and control of all relevant processes and activities, the adoption of full assessment procedures, such as environmental impact assessment, for new projects including the installation of new equipment, and the setting of performance criteria, which the controls will measure performance against.

Verification, measurement and testing procedures are to be designed and implemented within or in support of the controls, and these are to include records of results. Each *relevant* activity must be verified or measured by identification and documentation, specified verification procedures, levels of satisfaction, and procedures for malfunction.

For the reader attempting to relate this to a real enterprise, this means identifying those processes and activities, obvious from the Register of Regulations, or from policy or industry grouping standards, and ensuring that both procedures to control them and measurements are in place. How to manage this in a practical manner will be made clearer in later chapters. As with the quality standard, procedures for action in the case of non-conformance are required. These have to determine causes, have a plan for corrective action, give rise to future preventive actions, be subject to control, and be recorded.

6.9   Environmental management records

A deceptively brief paragraph covers environmental management records, but the environment standard, like the quality standard, must be supported throughout by records or documentation. The main documents have already been referred to: operational procedures, registers, environment manual, but in addition, all of these, plus contractor, procurement, audits and training records, must be controlled. This means indexing, controlling revisions, numbers of copies, security of the master, filing, and disposition. The standard also requires that they be *available*, both within the organization and to interested parties.

6.10   Audits and reviews

An audit plan must be devised to measure the effectiveness of the environment management system. Specific areas to be audited are the organizational structures themselves, to ensure that there is a relevant organization allocating responsibility, for example, working procedures, capacity and equipment, documentation, and, of course, the required performance results.

Frequencies must be specified as well as responsibilities and the 'qualifications' or independent status of the auditors. Methods of conducting audits and reporting procedures are also specified. These will be examined in more detail later as the implementation procedures are examined.

7

ISO 9000 as a Basis for BS 7750

The practical realities of adopting BS 7750 before achieving ISO 9000 are discussed in Section 7.4, but it is not recommended that anyone embarking upon the environment standard should do so without knowing about and understanding ISO 9000. This book is based on the author's previous book, *ISO 9000*, also published by Gower. A thorough understanding of ISO 9000 is the most valuable foundation for understanding the meaning of the environment standard and knowing how to implement it.

7.1 Description of BS 7750 – ISO 9000

BS 7750 is being described as a revolutionary new standard. This is a fairly safe assumption, as, by now, it has become apparent that its predecessor, BS 5750 was a truly revolutionary standard. Apart from standards for products and processes, this was the first internationally recognized and accepted standard for a management system, in this case a quality management system. This is not to say that there were not such internationally adopted systems. They were there in numerous formats in what were called 'vendor assessments'. A vendor assessment was the audit carried out by the sophisticated buyer on the aspiring supplier, and these rigid assessments had become common in dealing with military, nuclear, aerospace, automotive and electronics industries. In the drive for unification within Europe, it became obvious that such a proliferation of vendor assessment schemes made a unified approach impossible, so that a truly open market demanded a universally accepted quality management system, or at least a generic model.

BS 5750 became just that, a generic quality management system, which could be customized by specific sectors, but which was a universal model as far as basic product manufacturing was concerned. Soon the generic model was even adapted to produce a standard for services.

It was fortunate for the EC that the European standards body, CEN, was already in place as an existing unified organization at the beginnings of the Community. The members of CEN are all the national European standards bodies, such as BSI. These, in turn, also sit on the committees of ISO, the Geneva-based International Organization for Standardization. It was inevitable therefore that ISO would base an international standard, ISO 9000, on BS 5750, so that the latter became, or was equivalent to, ISO 9000. The next step was for the EC to ask CEN to adopt ISO 9000 for the Single Market as the agreed or 'harmonized' quality management standard. This provoked the epidemic-like spread of ISO 9000, the official European number, or European Norm, of which is EN 29000.

We are only now beginning to appreciate the meaning of ISO 9000. For the first time companies can practise and demonstrate real quality of operations and of product and service, and be supported in these claims by independent third party corroboration, in the form of certification, or (as it is called) 'registration'. One effect has been the dropping of spurious quality claims in advertising. If you really have quality, you should have ISO 9000.

Now, one can assert not only that one operates in a quality manner producing quality products, on a continuous basis, but also that it is not done in a manner injurious to the environment, or unsafe to staff or the community. This is truly revolutionary and, for the first time, allows industries to demonstrate that they can operate in a truly facilitative manner, satisfying both their customers and the environment in which they operate.

BS 7750 thus goes hand in hand with BS 5750, creating with it the ultimate standard, quality and clean and green.

7.2   Companies already ISO 9000 registered

These companies are in the best position to implement the environment standard, but for readers not yet familiar with ISO 9000 the following short description of the significance of achieving it is given in the example of a crystal manufacturer.

A crystal manufacturer who has achieved ISO 9000 has installed a quality management system, which is regularly audited by an independent certification agency. The 'registration' to ISO 9000 assures customers that the beautiful crystal products being produced for their stores worldwide will continue to be produced to the same specified quality, and that all the quality control procedures and their correct management mechanisms will stay in place.

Now, without BS 7750, that crystal factory could go on supplying the shops of the world with guaranteed quality products, but could do so with belching furnaces, black emissions, poisonous discharges to the surrounding waters, and solid toxic waste. It could also be dangerous to its workers, a threat to its neighbours, and an eyesore.

ISO 9000 put in place all of the procedures needed to maintain the quality of the products and the integrity of the processes needed to maintain that quality. BS 7750 puts in place the procedures to ensure that the quality achieved and its supporting processes do not harm the environment. ISO 9000 deals with the processes which support quality; BS 7750 deals with the processes which support the elimination of damage to the environment. A number of the methods and controls are common to both.

The standard has both an Annex B and a Table B.1 showing the links between BS 7750 and BS 5750. Both may be useful and of some assistance to understanding the relationship between the two standards. As most readers, however, will be comparing the standard with ISO 9000 (remembering always that ISO 9000 and BS 5750 are now one and the same), the following comments will refer to the links between BS 7750 and an implemented ISO 9000 quality management system, not to the numbered sections of the standard.

Put at its simplest, what this means is that ISO 9000 holders should have their Quality Manuals in one hand and this chapter, or BS 7750, in the other, as the installed Quality Manual, rather than the paragraphs of ISO 9000, is what they will be drawing upon. First let us list the elements in BS 7750 which can be virtually copied and adapted from ISO 9000.

- Commitment and policy.
- Organization and personnel.
- Records and control of documentation.
- Audits and reviews.

Next, the sections which will require to be fully devised and written, but which can be based largely on ISO 9000 experiences.

In establishing the 'Environmental effects', attention must be paid to the following details already catered for in ISO 9000.

- Contract review or customer brief.
- Procurement.
- Handling and storage.
- Packaging and delivery.
- Servicing.

In establishing both the main environment management system and its operational controls, the following steps catered for in ISO 9000 can be used as inputs.

- Design control.
- Procurement.
- The operational or process controls or procedures.
- Inspection and test.
- Control of measurement and testing equipment.
- Control of non-conforming product.
- Handling, storage, packaging and delivery.
- Servicing.
- Statistical sampling.

Throughout the standard, reference is made to the similarities between dealing with a customer and satisfying the demands of the environment. Comparisons are made between drawing up a specification in discussions with a customer and deciding what measures are needed to satisfy the demands of a clean environment. The very legislation or regulations can be seen as parts of the specification.

7.3 Companies on the road to ISO 9000

Because of the market demands on companies to demonstrate conformance with ISO 9000, the only responsible answer here is first to go for the completion of ISO 9000 if one is a manufacturer under pressure from buyers who are demanding conformance. At the time of writing, service companies are still not under the same pressures; in their case one can say that the only reason they should not complete registration to ISO 9000, would be if they were under pressure to demonstrate compliance to environmental regulations. On the plus side, however, BS 7750 and the ISO 9000 services standard are very similar, so, if one had achieved the former very little further would be needed to claim the latter. The expression 'services standard' is used in its familiar sense, but the more legal description is that ISO 9004 Part 2 is used to implement ISO 9001 to achieve the services standard. For more information on this subject see the second edition of *ISO 9000*, published by Gower.

As the pressures to demonstrate compliance to environmental directives increase, we may find the environment standard being demanded first; this becomes more likely when we reflect that achieving the environment standard is virtually tantamount to achieving ISO 9000, so little is left to do to achieve the latter.

Where there are either strong environmental pressures, or where a clear marketing advantage in achieving the environmental standard is sought, all that needs consideration is the question of whether a company can afford the time and resources to appoint a person other than the busy quality manager, who will already be implementing ISO 9000, to develop BS 7750. The latter would be concerned with the separate issues of the Register of Regulations and the documentation and control of those activities relating to the environment. While the quality manager was involved with the production line's integrity of performance, the environment manager will be examining it and all of the inputs and outputs for environmental effects. In addition, the environment manager will be studying the drains and the chimney stacks.

As no clear cut lines can be drawn here, companies on the road to achieving ISO 9000 may also find the following section useful.

7.4 Companies not yet embarked upon ISO 9000

By the time this book is in print, the reader of this section must, if a manufacturer, be asked why he or she is not on the road to ISO 9000, as sophisticated buyers everywhere are asking for it. If not, however, whether a manufacturer or a service company, the intriguing question must be asked, which to go for first?

In an ideal world, the answer must be BS 7750 into which a quality management system, aimed at quality of product or service, will be embedded. This will be the ultimate standard. And this is a real possibility for service companies. As far as manufacturers are concerned, however, the demands of buyers are such that any delay in achieving ISO 9000 could result in loss of market share.

The recently opened-up public procurement market in the EC, for example, will be virtually impenetrable without ISO 9000, particularly as local standards cannot be used to exclude non-national EC suppliers. At the time of writing, demand for ISO 9000 from all of the sophisticated and hi-tech industries was widespread. It has become the norm in electronics and related industries.

At the same time it would be a smart move for a company whose competitors had achieved ISO 9000 registration to go one better and implement BS 7750, especially a company which badly needs to support its clean and green claims in the marketplace. This raises the interesting question of who needs the standard most and who will find it easiest to instal.

7.5 Practical realities of attempting BS 7750 first

Bearing in mind always that the question of what the buyer is demanding may have to be answered first, one of the practical realities of attempting BS 7750 first is that it will be much more easy for some companies than others. On the other hand, some companies will need it more than others for marketing or for good community relations.

Some examples may help here. A software company might have to pay attention only to the quality of its packaging material and instructions about the ultimate disposal of its diskettes. If its packaging supplier is already complying with regulations on the manufacture of packaging material, and a safe code for disposal of diskettes, or their re-cycling, exists, the software company could have an easy passage, with little to do other than demonstrate its awareness of the relevant directives and confor- mance with them, to achieve the standard. There is one exception to this, and it is that the software company could be worse off than a chemicals processor if its end products are not ergonomically sound and competently usable, and especially if confusion in their use could cause accidents.

A food and drinks company, especially in the dairy sector, could have an immense and even urgent need to demonstrate that it is clean and green, on its packages and over the media in advertising, but it will have a lot more work to do in the processing area than the software company. This extra work, however, can be rewarded by the fact that its green image is worth more in sales for it than is the image of the software company, whose customers (end users) may be less interested in its environmental status. The food products are more likely to come to the notice of major buyers, such as distributors or state procurement officers, who may, and probably will, begin to demand the environment standard. In addition, the software company can never gain as much advantage by employing a green image in the marketplace as a food and drinks company can.

A chemical company may have a great and urgent need for the standard to both satisfy existing regulations and demonstrate its neighbourliness within a community. It will be in one of two extreme situations: either a modern chemical plant, so well designed and controlled, and already so far advanced in environmental awareness and conformance that it will find the acquisition of the standard simple; or a plant with so many problems that only an immense amount of work, re-equipping and investment will bring it up to the standard. A consultant looking for a quick BS 7750 success would not be advised to tackle such a company; yet it could be a big customer because of its needs.

Power utilities are most interesting in relation to the standard. Natural gas suppliers have an immediate positive advantage. It is easy for them to be clean, and they can rapidly instal a system and acquire the standard, which will give them a market advantage. Power utilities which still have emissions in breach of regulations are in a most invidious position, for not only will they not achieve the standard until the emissions are eliminated, but they may indeed stop others achieving it. At the time of writing it is still not clear how far down the supply chain the certifying inspectors will go, but if the promise that audits will follow ISO 9000 routines, then energy suppliers and the effect that the production of their supplies has on the environment are a fundamental part of the procurement procedures being demanded by the standard. This could give rise to the extraordinary and even bizarre situation that unless a company could prove that it was taking only clean power off a national grid, it could not achieve the standard. It would have to try to make its power utility clean up all emissions. National power companies, not conforming to regulations, could find themselves being the cause of national industry not achieving the standard and, in this way, having a negative effect on national output and exports. This somewhat alarming picture must be moderated by the fact that the first version asks that suppliers be requested to comply with regulations: it does not demand that suppliers so comply, but what starts as a suggestion tends to become a regulation in environmental matters, and, in the case of ISO 9000, buyers are already expected to force conformance onto suppliers. One suspects that the initial soft approach of the environment standard in this respect is part of the growth process, and that once the standard begins an epidemic-like spread, suppliers may have to conform to stay in business.

To summarize, the practical realities of obtaining BS 7750 will be determined in the first instance by legal pressures and market advantage. Any company whose supplies could have a detrimental effect on the environment and, in particular, could be legally exposed by non-conforming activities may need the environment standard immediately, and also ISO 9000, to demonstrate total management competence and care. Other companies attracted by the ease with which they can implement the standard and by the consequent rewards in good public relations will also go for it rapidly.

If the standard is the success that many hope it will be, it may replace the current ISO 9000 altogether, or ISO 9000 may simply expand to embrace it.

8

The Elements Involved

This book is intended to serve both as a guide to BS 7750 and as a methodology for implementing the project of installing an environment management system to meet the standard. This chapter is designed to help understand the project in its overall sense and is aimed at the project co-ordinator or leader or project team. It is part of the section on definition and scope, but comes before later sections of the book which deal with the installation stage. The chapter attempts to describe the more important elements involved in the project.

8.1 The main elements

The main elements may be described as initial appraisal, commitment, preparatory review, establishment of issues, design of system, implementation. These cannot be set absolutely in a fixed order for all organizations; the chief executive might, for example, impose the standard, which would move commitment to the top of the elements; a potential projects leader on the other hand might have to carry out early appraisals, perform cost/ benefit analyses and make a proposal.

From the project point of view, however, the steps are as follows: to find out, propose and get a commitment; do a quick first review to see what particular regulations apply; to carry out a detailed investigation of the real issues; and to design and implement the system. All these are dealt with in the following chapters in the language and format of the standard, because it is in that language and format that the certifying inspector will attempt to do the audit.

8.2   The main documented steps

Another useful way of viewing the project is in terms of the main documented steps, as the major documents formalize each main element or stage. These are:

- Initial plan.
- Project plan.
- Preparatory review.
- Inventory of regulations (called Register of Regulations).
- Inventory of environmental effects (called Environmental Effects Register).
- The Environment Management Manual.
- The Environment Management Programme – this is the full system with all documentation for procedures and controls.
- Training.
- Pre-registration audit.
- Application for certification.

8.3   The planning phase

The planning phase is covered in later chapters and involves forming initial plans and a project plan, getting management commitment and constructing policy and an organization. The main documents involved here are the plans themselves, the report on the results of the Preparatory Review, the Register of Regulations and the Register of Environmental

Effects. If one is in the happy position of also planning a new facility or organization, this is also an excellent time to carry out an environmental impact assessment, as mentioned in Chapter 3, and to use the design phase of products, processes, or services to incorporate environmental protection measures.

8.4 The implementation phase

The implementation phase entails the construction of an Environment Management System and puts into place the controls and documentation needed to ensure that the activities of the organization meet the targets set by both the Register of Regulations and company policy. The latter may in the first and subsequent years after achieving the standard improve year by year on the demands set by the regulations. The Register of Environmental Effects forms the basis of the Environment Management System, as it dictates what regulations and targets are relevant to each organization. This phase will also involve staff training, both in environmental matters and in health and safety regulations.

8.5 The certification phase

The certification phase is largely out of the hands of the organization except for the application and other steps covered in Chapter 14. A pre-registration audit will, of course, be possible, using either internal staff or an outside consultant.

8.6 The main documents

Although there are a large number of documents called for, there are three main books or registers, and these are:

- The Register of Regulations.
- The Register of Environmental Effects.
- The Environment Management Manual.

All three are discussed in detail later and samples given.

The next most important group of documents is the operating procedures. These will vary from organization to organization, and will, for example, be very different in a processing industry from a service industry. These should be designed and filed in separate controlled binders and cross-referenced to the Environment Management Manual. It should be specifically noted how amendments and updates are to be dealt with.

Other documents include the policy statement on the environment, a policy statement on safety, records of audits, and records of training.

Although the plans are vital initial documents, they will not form part of the continuing documentation which needs to be maintained and controlled, although the inspectors will in the first year expect to see evidence of the Preparatory Review, and a report on it.

8.7 Steps, documents, systems

Some of the activities needed to implement the project are plans, some are steps in a one-off implementation project, and some are part of a continuing system. The results of all the initial steps and the implementation phase are the construction and implementation of the Environment Management Programme, which perhaps might have been better called the Environment Management System.

This is the continuing system based on a Register of Regulations, which may evolve and need updating. This means that the initial Register of Environmental Effects will become less important as time passes, even though the standard demands it

8.8 The up and running system

Once the implementation stage is complete and the standard achieved, the situation becomes much more clear and is shown diagrammatically in Figure 8.1. It will simply be an Environment Management System based on environmental and health and safety regulations, which is fully managed, documented, and audited.

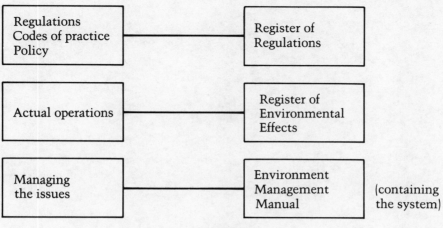

FIGURE 8.1

8.9 The personnel element

The standard places great emphasis on personnel involvement and training. ISO 9000 has taught us that these were fundamental to its success. It also revealed the huge benefits from savings, and increase of staff morale, after the achievement of ISO 9000. What applied to ISO 9000 will apply as much, and more so, in the case of the environment standard.

The personnel elements are discussed both in Chapter 11 (Policy and Commitment) and Chapter 18 (Staff Training).

8.10 Other parties

The most important other party in achieving the standard is the supplier to the organization, except where the customer has first demanded it. Both supplier and customer are technical partners in the standard, but beyond them is the community at large at whom the standard is also directed. Other important audiences are potential employees, prospects, regulatory bodies and the media.

9

Costs and Benefits

The subject of costs is strikingly different from that of quality costs, where all of the costs involved apply to the costs of the quality control system and to the real costs of poor work, re-work and scrap. In the case of environmental costs, there is cost both to the organization and to the environment itself. In this chapter, however, we will consider costs and benefits only within the confines of the organization.

9.1   What the standard says about costs

Costs are mentioned in the standard in the context of providing sufficient resources, as a possible guide in setting environmental objectives and targets, and as a yardstick in the preparatory environmental review. Although the last item gives some insight into possible benefits, including operational cost reduction, in general the standard has little to say about costs.

9.2 Project costs

The project costs will be those incurred in designing and implementing the Environment Management System and applying for certification. These costs will include the salaries of staff, fees to consultants, or other outside experts, and time spent on staff training. Whether or not one uses full-time staff will depend upon the size of the organization, the nature of the operations, and the existing sophistication of both the quality management system and environmental control systems.

It would seem that a small to medium-sized enterprise of up to 100 to 250 staff could handle the task using existing staff, in the case of non-process industries, or of those process industries with highly developed environmental control mechanisms. Large organizations, particularly those in process industries, or with many divisions, will need full-time staff and even teams in some cases.

The above remarks refer to staff and related costs only. If, however, only new or adapted processes and equipment can meet the standard, such as the scrubbing of stack emissions, there may be huge costs, in some cases more than a company can bear. What will be decided here is the question of whether the enterprise is profitable only at the expense of the environment. In very bad cases, such as a mining or building company extracting gravel from a river bed or from sea beaches, society and legislators may pass judgement on such operations long before a standard becomes relevant to them. Many companies will find that they need to change operations and implement the standard, as only both will satisfy the regulatory authorities.

9.3 Running costs

The running costs will be the annual costs of operating the Environment Management System, including those of audits, upgrades and annual inspections. The other costs, which will be insignificant or major depending on the extent to which processing interacts with the environment, will relate to the costs of controlling, neutralizing, re-cycling disposing or otherwise treating the outputs of processes, including the costs of hardware and software.

9.4 The benefits

All these costs will be reduced or offset by the benefits, to an extent which will be decided by a cost–benefit analysis. Some of them, however, may be quite substantial. We will look at them now in terms of direct and indirect benefits.

Potential direct benefits

- Reduction in resource consumption
  (a) raw materials
  (b) energy.
- Reduction in scrap or waste.
- Reduction in complaints and follow-up.
- Avoidance of accidents or emergencies.
- Avoidance of claims.
- Avoidance of fines and penalties.
- Avoidance of personal liability.

Potential indirect benefits

- Enhanced corporate image.
- Enhanced marketing capabilities.
- Improved staff morale.
- Better customer relations.
- Better community relations.

9.5 Statutory and marketing demands

Statutory and marketing demands are not, strictly speaking, benefits, though many a cost–benefit analysis concludes with the management decision that they have to be met anyway. The environmental legislation, and the health and safety regulations, make demands that simply must be met, regardless of costs, while many marketing demands, especially those of major customers or public procurement buyers are tantamount to being mandatory. A good way to analyze costs/benefits therefore is to set both statutory and marketing requirements as the benchmark against which the costs and benefits of implementing an Environment Management System can be measured at the minimum required level, and to simulate costs and benefits above that benchmark.

A growing number of companies, however, have a different primary motivation, or strategic policy, and this is reduction of exposure in the event of non-conformance of product, service or process. This could include something as simple as a careless action on the part of a tanker driver. Protection against potential charges of negligence, both personal and corporate, demands a demonstrable system of management control, and BS 7750 is a standard which offers verifiable proof of such a system.

9.6   Cost–benefit analysis

As a target for both the achievement of the standard and the first year of operating to the new standard, a company can specify a set of objectives which meet statutory and marketing requirements. There will be some fixed and variable costs and the benefits will vary also. There will also be one-off costs.

Fixed costs (approximately fixed)

- Designing and implementing the Environment Management System.
- Operating and auditing the system.
- Inspections and certification.
- Annual operating costs.

One-off costs

- Designing, purchasing and installing control equipment.
- Designing control systems including software.
- Disposing of accumulated waste.

Variable costs

- Operating control equipment.
- Purchase of chemicals such as cleaning materials.
- Degree to which controls are exercised or outputs reduced above statutory requirements.

PART III

Installing the Standard

10

The Initial Steps

One valuable lesson from ISO 9000 is that trying to find a way to start the project of implementing a management system standard may be the most difficult task. The main reason is the difficulty in getting the attention of a decision-maker near, or at, the top. Indeed, it has to be the chief executive of small to medium-sized companies and a senior vice-president in large companies or corporations. Once the process begins in sectors such as manufacturing or services, then the market itself begins to demand it, but, at the very beginning, unless some senior manager is well-briefed, or far-sighted, it may be difficult to get the project off the ground.

There are three main reasons for a company to decide to go for the environmental standard. These are: legislation, market demand and marketing advantage. If the first two are absent, or not yet fully apparent, the third will be the catalyst. As time passes, however, the first two will become dominant.

10.1 Starting out

The internal impetus needs to come from the quality manager, or from marketing or public relations. It may also come from some knowledgeable engineer who has been responsible for implementing environmental controls to meet existing regulations.

Unfortunately, this standard creates an organizational problem not encountered under the ISO 9000 regime. Where one simply appointed a quality manager, before, now the question is do we need two 'quality' managers (one for the quality of the product or service and one for environmental operations)? A possible answer will be spelt out later in Sections 11.4 and 13.9, and it will depend upon both the size of the organization and the existing quality status.

This question, however, could have a profound effect on how one starts out, as the possibility of potential conflict or competition arises. It is essential to sort out this question along the lines discussed in Section 11.4 before embarking on the procedure for getting started. Getting acquainted with BS 7750 is the next step listed, but it is also fundamental to starting out properly, as one needs to be briefed thoroughly before approaching top management. As the standard spreads, more and more consultants will move into the propaganda campaign needed here, influencing top management directly, or via their public relations and marketing functions.

10.2   Getting acquainted with BS 7750

The obvious way is, of course, to purchase *BS 7750 Environmental Management Systems* (1992) from BSI, Linford Wood, Milton Keynes, UK, MK14 GLE. While this is a beautifully written document, it is very general, running to a mere twenty pages, but covering a multitude of topics. Read, together with this book, however, and related to one's own organization, it will make a lot of sense. It will make particularly powerful and compelling reading for quality managers who have already implemented ISO 9000.

Once both the standard and book have been read, the next obvious step is to study the two sets of regulations – the environmental regulations and the health and safety regulations. This is very important, as the moment one approaches top management, one will be asked the question: How will it affect us? It would be embarrassing not to know.

Finally, before making the approach to senior management one should have at least a brief chat with the engineers already, or about to be, involved with implementing controls, to ascertain what level of extra investment would be involved in meeting any regulations not yet complied with.

The simplest form of approach to top management is an internal letter accompanied by a one page statement about the standard. Here are suggested samples.

To: Chief Executive

From: Quality Manager

I wish to bring to your attention the existence of an Environmental Management Systems Standard published by BSI, entitled BS 7750. A short description is attached.

I strongly recommend that we should study and implement this standard, and I will be happy to supply more information if requested to do so.

Regards,

Quality Manager

The above letter should be accompanied by an explanatory statement, such as the one shown next.

Accompanying statement

Background to the new environment standard
A revolutionary new management standard was published in 1992 by the British Standards Institution (BSI), entitled *BS 7750 Environmental Management Systems*, a standard for environmental control within the manufacturing and services sectors. BSI was first with BS 5750, which was the model for ISO 9000, and has now repeated that pioneering effort with a new environment standard. Both the EC Eco-Management regulation and a new ISO standard are expected either to mirror or adopt BS 7750. The EC has already adopted ISO 9000 (EN 29000) based on BS 5750, and a similar course of action is now under way with this new standard. Since 1992, BS 7750 is available for actual implementation by companies, and a certification scheme is in place (or will shortly be, depending on the country, or certification may be obtained elsewhere), which also meets the demands of the EC Eco-Management regulation. Buyers, particularly those in public

procurement, will now begin to look for the standard.

This revolutionary new standard means that companies can demonstrate both quality management and environmental care, each supported by third party corroboration. It is virtually certain therefore that unsupported 'green' claims will be proscribed. For the first time industry can demonstrate that it is truly facilitative towards the community and the environment. It has already been accepted that companies adopting the environment standard will aggressively market it.

Companies such as ours that already have ISO 9000 will find the new standard easy to adopt, as it is an environment version of ISO 9000; however, the environment standard is already being talked about as the 'ultimate standard', which at this stage seems to suggest that it will be considered to be as good as both standards.

This is also a mechanism for incorporating the compulsory Health and Safety regulations in force since the end of 1992, as the standard covers these as environmental matters affecting employees. This, on its own, would appear to justify adoption of the standard, as it is difficult for us to manage the health and safety regulations without a formal control mechanism.

Finally, BS 7750 would enable us to demonstrate conformance to all the regulatory instruments and codes of good practice within our industry, which would have vital implications for us in cases of potential liability for incidents involving our products or services in relation to the environment, staff or the community.

Opportunity for our company

It is possible for us to become one of the first companies, and probably the first of its class, to achieve both the new BS 7750 Environment Standard and the EC Eco-Management regulation. This would allow us to claim credit for both, to display the BS 7750 and Eco-Management logos, and to earn a considerable amount of good publicity for our environmental policies and practices. On the practical front, at this stage there do not appear to be any formidable obstacles; on the contrary the company seems to be well-advanced on the three broad, relevant matters of environment, efficient and safe product usage by customers and public safety. In the case of our new plant, still under construction at Highway Seven, there is the enormous extra advantage that we are still at the pre-implementation phase, for example, the ISO 9000 system for the plant has not yet been planned.

There is a further potential bonus for us in that the BS standard is a convenient mechanism for meeting the mandatory requirements for a number of new health and safety regulations which have become a legal requirement this year and which we have to cater for anyway.

If one is a consultant approaching companies, the following letter together with the statement is recommended.

Dear (Chief Executive)

You may be interested to learn that there is now an international Environment Standard, BS 7750, published by BSI in 1992. This is being followed by both a new EC directive on the environmental auditing of industry and by a new ISO standard. It will be the ultimate standard, and any company involved in any way with the environment will find it essential to conform to it.

Of special interest to you is that you can achieve it both with or without ISO 9000. Having ISO 9000, or being on the way, is an advantage. There will be a great interest in companies being first in their fields as this will give them a market advantage over the competition. Any company operating in an environmental area such as yours will want to be amongst the first, especially those with environmental names.

On a separate sheet I have attached more details about this new standard. I have a number of clients with whom I am working on the installation of this new standard and would be delighted to include you amongst them. I believe that your company is in a position to profit greatly from achieving the standard, and that it is also well-placed to implement it.

I shall telephone you next week when you have had time to consider this matter.

Yours sincerely,

Consultant

10.3   Appraising the situation

Annex A to the standard has an excellent paragraph (A.1.2) titled 'Preparatory environmental review', which is aimed at organizations with no existing formal environmental management system, but which also seems suitable for all organizations to help establish the status of their conformance to regulations and how far they have to go to implement the standard. The approach is very like that of a preliminary audit before implementing a quality management system. At this stage all that should be necessary to achieve management approval is to inform management that a preparatory environmental review will be needed at the commencement of implementation. Section 12.4 describes where this fits in. The contacts with appropriate engineers should ensure that one has some knowledge at this stage of how far an organization will have to go to devise controls.

What a brief reading of this section will reveal, however, is how much work has to be done, not just to instal operational controls but to implement the control and paperwork needed for the standard, allowing the proposer to estimate the number of man months or years involved.

A word of caution, however. It is not until one actually carries out the preparatory review, which cannot be done without the go-ahead of management, that one can know how much work will be involved. In the initial approach, therefore, one should include this caveat. The very least that can be expected is authorization to proceed with the preparatory review. Of course, one may also be asked how big a job that is and how long it will take. It needs to cover four areas:

- The legal requirements.
- An understanding of the organization's operational effect on the environment.
- A review or audit of existing practices.
- A review of past history in this respect.

10.4   Preliminary audit

The preliminary audit should also wait until the implementation phase begins, but a little knowledge of what is involved may help with the initial proposal. This is where an actual audit is carried out as part of the preparatory review involving questionnaires, interviews, shopfloor inspections and so on.

All one needs to know at this stage is what the major issues are for the company, to be aware of the main regulations, and to be in a position to do a scoping study to establish the size of the project.

10.5   The proposal

After the initial submission by letter, or in person, a proposal may be demanded. This can be based on the following outline.

1. Background.
2. What it means for our company.
3. Relationship with ISO 9000.
4. What are our environmental issues?
5. How good or bad do we appear to be?
6. The need for a preparatory review and audit.
7. Recommendations.

The proposal is pitched to include what will be known at this stage without trying to include information which cannot be fully finalized without a more detailed review and audit. An outside consultant would be expected to produce less at this stage. A consultant can attempt one of two broad approaches. Firstly, try to get a company to agree to a scoping study or preliminary audit, which would reveal the extent of the work necessary to implement the system; secondly, preferable from a business point of view, try to get the company to commit itself to the system and embark upon implementation. The second course has one danger for both customer and consultant, more important perhaps to the latter, namely, the difficulty of estimating the job without a preparatory review or scoping study, and the consequences of incorrect financial estimation, including that for the consulting fee.

A useful approach is to estimate times and cost only to a certain level and then to re-negotiate. An example follows on the next page.

Dear (Chief Executive)

Thank you for the enthusiastic session with you and your team on Monday last.

My proposal is that you make a policy decision to implement BS 7750, the new environmental management standard, in an attempt to be the first to achieve it in your group, and possibly the first in your business worldwide.

If you agree, I would be pleased to assist in the next steps, which would be:

1. A plan of action produced within one week of go-ahead.
2. An inventory of the relevant environmental directives, as complete as possible within one month.
3. The development of the Environment Management Manual, outline with draft contents within three months.
4. Full documentation, controls and implementation on site, six to twelve months from now, assuming no hold-ups due to technical environmental problems which need solution.
5. Application for registration on completion of the above.

I would be happy to supply most or all of the material up to item 3 above — that is, up to and including the development of an outline Environment Management Manual, within the three month framework, for a fee to be agreed with you, with an upper ceiling, based on the amount of work and time involved.

When it comes to the full documentation and implementation phase, we can negotiate my continued involvement should you still find it beneficial.

I can quote you formally or discuss it over the telephone, just as you require.

I look forward to hearing from you.

Yours sincerely,

Consultant

10.6 Plan within the proposal

If the proposal is accepted one may then produce an outline plan. We are still discussing proposals here, which may have to involve some projected plans. The project plans will be discussed in Chapter 12.

10.6.1 Outline plan of action (within one week of go-ahead)

The plan would contain:

1. List of initial steps to be taken.
2. Outline of all steps from now to registration.
3. Approximate times.

As the job proceeds beyond the initial steps, the plan may need revisions and updating.

4. Inventory of relevant environmental directives (within one month), containing a list of national, international, and EC regulations which affect the company. This will need verification by staff before it is complete.
5. Development of the Environment Management Manual (outline with draft contents within three months) containing:
   - An outline generic manual customized to the company as much as possible at this stage, i.e. including the processes relevant to the standard.
   - Cross-reference to registers of regulations and issues.
   - ISO 9000 material which can be used incorporated into the manual.
   - Outline of the areas of production affected by the standard with appropriate spaces for the procedures which may still have to be developed (*viz* dealing with toxic waste).
6. Full documentation, controls and implementation on site (six to twelve months from start), including:
   - All procedures identified, and controls specified.
   - Full documentation.
   - Implementation on shopfloor.

- Vendor assessment of supplier conformance to any directive demanding attention.
- Design considerations.
- Housekeeping.
- Inspection and testing procedures.
- Design of audits.
- Organization changes.
7. Training
8. Application for registration, including:
    - Pre-registration audits.
    - Liaison with BSI.

10.7 Making the commitment

Establishing a commitment is fundamental to the initial steps and details of this are handled in the next chapter. A number of companies will find this easier than others, especially ISO 9000 holders, who may be looking for mechanisms for continuous improvement and the environment standard is an excellent target for that. Also high on the list are those companies facing potential legal liability or stringent environmeenntal regulations, including regulations for the health and safety of their workers. A great stimulus is, however, market demand, and all it may take is one large customer asking for it to galvanize management into action. The proposer will weigh up all of these factors as well as the marketing and PR advantage. Once the commitment is made, it must then be recorded along the lines laid down in the following chapter.

10.8 Making a plan

There will, in fact, be several plans and one big one. Initially thumbnail plans are needed for the proposal to management and for the preparatory review or preliminary audit. The big plan, however, will be for the implementation of the Environment Management System itself. Although, strictly speaking, this is not the 'standard', common usage of the process refers to this project as the 'implementation of the standard'. What is meant is the system controls and reviews which lead to successful certification and the achievement of the standard. The plan is so important that a sample is given in Chapter 12.

A summary of the steps to this starting point is as follows:

Identifying or responding to the initial catalyst

- Legal liability.
- Legislation.
- Market demand.

Preliminary appraisal of situation

- Acquaintance with BS 7750.
- Evaluating company status.

Proposal to management

- Initial commitment.
- Preparing the plan.

11

Policy and Commitment

In Section 6.4.2, policy was briefly examined as were organization and personnel in Section 6.4.3. In the standard itself, policy is covered in fair detail in both Section 4.2 in the body of the document and in Annex A, Section A2.

11.1 Definition

In the definitions given by the standard, environmental policy is defined as 'A public statement of the intentions and principles of action of the organization regarding its environmental effects, giving rise to its objectives and targets.' This rather legal statement can be interpreted as a published policy statement announcing the intentions of the company to implement an environmental management system and to have it monitored for conformance to both its own stated targets of meeting the demands of the regulations, and its own environmental targets. It also implies in this instance that the company will seek certification to BS 7750 or a similar standard. A public statement means hanging a framed policy statement on the walls of the company's lobby and including it as one of the first paragraphs in the Environment Manual. The ISO 9000 policy is already expressed in a framed statement and in a first paragraph in the Quality Manual. It is now clear that at least two expressed policy statements are expected to be supported by framed certificates – ISO 9000 and BS 7750, upon their issuance from the certifying authority.

11.2   Objectives

The standard defines the environmental objectives as 'The goals, in terms of environmental performance, which an organization sets itself to achieve, and which should be quantified wherever practicable.'

As we saw earlier, what is expected here is the meeting of all legislative and regulatory requirements, at the very least, and other targets as they are set by policy. Once again, it may be helpful to caution readers against over-specifying, as the inspection agency will hold one to these high objectives. As a commitment to continual improvement is also expected, this should help to avoid setting too high a target initially. Perhaps one should begin with the basic regulations, and then improve step by step each year.

11.3   Scope of policy

Annex A reveals that the scope of the policy is very wide indeed. It wants the policy stated from the top with a commitment to support it also. It wants it to be consistent with other policies, such as that for meeting the health and safety regulations, the quality policy and other relevant policies.

It also wants to see expressed in the policy the intention of management to implement the system and controls necessary to support the policy and make it meaningful. This includes the meeting of, or doing better than, regulatory and other relevant requirements. Finally, it asks that a system of continual improvement be put in place, and that the policy be made available, suggesting publication of it and its display in a public place.

The above concerns the formulation of policy fundamental to the day to day operations of the organization, but the standard goes much further, suggesting that a company may wish to question where its funds are invested and ask if they are used in an environmentally responsible manner. This is most interesting for while on the one hand it is not mandatory that a company go so far with its policy, on the other, both increased environmental consciousness and the need to avoid possible damaging publicity will influence companies to ensure that all of their activities are environmentally sound.

While the standard gives considerable scope to how the policy will be expressed, it suggests that the following commitments will need to be made: a reduction in consumption of materials and energy, and a reduction in waste, the reduction and elimination of polluting emissions and discharges, the minimization of detrimental environmental effects at design wherever possible, the control of the environmental effects of raw material sourcing, and the use of strategic planning to minimize any adverse affects of new developments.

The control of the environmental effects of raw material sourcing is the only item of procurement mentioned under policy, although procurement itself comes up in the standard. In brackets after this sentence, habitats, wildlife and flora and landscape are mentioned. This appears to be a straightforward message that an organization should know how and where its raw materials are being mined, or harvested. This brings to mind timber, paper, board, minerals of all kinds and fuels.

11.4 Organization

Organization and personnel are covered in two sections of the standard, in the main body and in Annex A. This is modelled almost exactly along ISO 9000 lines. In effect, it asks for a clearly defined organization with a named management representative. This is the equivalent of the quality manager, who can still be a quality manager or a separate environment manager. The standard demands that whether or not she or he has other responsibilities, quality, for example, or engineering, that defined authority and responsibility be given to the role to ensure the implementation of the standard.

In this section it is also laid down that all personnel be informed and where necessary trained in the meaning of the standard, as well as the potential environmental consequences of their actions.

Annex A goes into more detail about the environment manager, called 'the management representative'. It requires that the environment manager have sufficient knowledge of both the activities of the company and of the environmental issues to do the job properly. In all cases with this manager, or with 'deputies' appointed for specific jobs, there should be no conflicts of interest – in other words production or other interests must never be allowed to over-ride environmental interests. The environment manager also has a responsibility to monitor evolving environmental legislation and regulations.

In an interesting paragraph the standard also demands that the environment manager develop and implement environmental management systems for each functional area, activity or process, such as product and process design, planning and development, safety, purchasing, packaging, production, distribution and even site services and facilities.

This section also calls for training for executives, all operating personnel and new recruits. It gets quite specific at this point, mentioning introductory and refresher programmes, recognition of performance, encouragement of employee suggestions and participation in environmental initiatives.

11.5 Sample policy statement

The policy statement needs to be framed and hung in the organization's front lobby and recorded at the beginning of the Environment Manual. Here is a sample statement.

> On 12 May 1994, the management of ABZ Corporation made the following policy commitment.
>
> 'It is the policy of the ABZ Corporation to manufacture and deliver products and services in a manner that is not detrimental to the environment or to the health and safety of personnel inside and outside our production facilities. It is also our policy to implement an environmental management system, which sets our environmental targets for all of our relevant activities. These targets and the results of our environmental management system will be made available to interested parties upon request. It is also our policy to support these targets by obtaining third party corroboration of the success of this system in the form of certification to BS 7750.'

11.6 The project manager

Regardless of who will be the environment manager, quality manager, engineer or other, the project of implementing the system to meet the standard is such that during the implementation stage it requires a project manager. This may be the person who will also be the continuing 'management representative' – the environment or quality manager. This question, unfortunately, will not be resolved during the writing of this book, as the standard is in its early stages of implementation, but on balance it would appear that the quality manager should assume total command of both the environment, including health and safety, and quality, except perhaps during the implementation phase in those process industries where a specialized engineer needs to be the project manager. Once the project is up and running, the quality manager can then take over. This means that the environment and health and safety are all seen as 'quality' issues, and this is certainly supported by the latest versions of ISO 9000.

At the point where the policy statement is written down, the project manager should be appointed; indeed, the writing of the statement may be his or her first task. At this point also a project team needs to be assembled. This could consist of the quality manager, an engineer, possibly the human resources manager, because of the degree of training involved, and the staff health and safety issues, and perhaps an outside consultant. The last is particularly relevant where the outside consultant has introduced the concept of the standard in the first instance, or where one has been brought in specifically for the task.

11.7 Staff training

So much emphasis is placed on the subject of training in the standard that a separate chapter, Chapter 18, is devoted to it.

11.8   Commitment and morale

Commitment and morale may be the most important argument in the whole process for a single committed person trying to get the organization to make a commitment. The step, once taken, may have a profound and beneficial effect on the morale of the whole organization.

As we saw earlier, any one or more of a number of motivations may inspire or galvanize an organization into taking action on the environment – potential legal liability, regulations, demands from buyers, competition, marketing advantage, public relations and so on. Also, as we have already noted, ISO 9000 bearers together with all 'world class' operators need a continuously evolving target to aim for and the environment standard is an excellent target for this purpose.

Possibly the best and most worthwhile motivation is for the staff of an organization to be able to say that they work for an organization that demonstrates that it cares for its environment, not in unsupported claims, but in a verifiable, independent third party corroborated manner to an internationally-recognized standard. All ISO 9000 achievers were rewarded by improved staff morale. The rewards in morale for a commitment to the environment standard may be even greater. The companies with both standards can point with pride to quality products or services produced and delivered in an environmentally caring manner. This is the ultimate investment in staff morale, the economic benefits of which may be very great indeed.

Finally, this may be the one time that an outside consultant is most useful. If the insider fears that making a case for the environmental standard may be misconstrued as power play or as seeking self-promotion, or, if an insider simply has not got the authority or status to make the case, or to voice fears about the possible future consequences of not having an environmental standard, an outside consultant may be in the best position to make the case.

11.9   Relationships

The project manager and the subsequent environment or quality manager must report to the chief executive and be given full responsibility for the environment standard, with no strings attached, such as being compromised by production or other operational considerations. The independent inspectors for the certifying authority will look for independence, authority and defined responsibilities.

A system of part-time deputies can be set up using persons with other line responsibilities within the key function, such as production, procurement, design and so on, and these will have a functional reporting role to the environment manager on matters concerning the standard.

Relationships are most important within multinational companies. Those with manufacturing facilities in Europe, particularly in the UK, will be likely to be the first to become aware of the standard. As with the growth of ISO 9000, these companies will achieve the standard first, and then help to propagate it to other sister companies worldwide. The experience with ISO 9000, however, has since revealed that a decision to go for such a standard throughout a multinational organization should be made at the highest strategic level. One or more plants should be chosen as a prototype or reference site and the standard implemented there in the first instance. Once successfully achieved there, rapid transfer can be made to other facilities internationally.

Other key relationships will be between suppliers and major equipment manufacturers or large service providers, and those between world-class suppliers and their customers. Most of the early activity will occur with major manufacturers passing down the standard onto their sub-suppliers, or 'vendors', but some sophisticated world-class component suppliers may also pass it up the supply chain to their customers.

12

Making the Plans

This chapter applies to the project phase, when one is designing the Environment Management System to be implemented to meet the requirements of the standard. The chapter also applies to all stages of the project from initial conception to certification. There are two basic plans, the initial plan and the overall project plan. In complex industries, such as process, especially where new procedures are required, other detailed plans will also be required, all of them subject to the overall project plan.

12.1 The initial plan

It will be useful to devise an initial plan to get the project up and running. There were two main reasons for this: firstly, some initial steps are often needed before obtaining the necessary management commitment; secondly, the project really has to be well under way before a full project plan can be embarked upon. It may not be possible to draw up a realistic project plan until after the preparatory review; even then the plan must be capable of adaptation as required.

The initial plan needs to contain the following main elements: first project planning meeting, appointment of project team, policy commitment, identification of first tasks, assignment of first tasks, the planning and carrying out of the Preparatory Review, and the commencement of the drawing up of the main project plan. Also useful at this stage are three outline checklists covering all the known steps to certification, the possible environmental issues, and the possible health and safety issues.

The initial plan can be demonstrated as follows:

Task	Assigned to	Date complete
First planning meeting	JK	1 June
Set up project team		1 June
Policy commitment		1 June
Identify first tasks		1 June
Assign first tasks		1 June
Preparatory review		30 August
Project plan (commence drawing up)		6 June

The useful lists are:

List of possible environmental interactions
> Planning
> Emissions
> Effluent discharges
> Solid waste
> Toxic waste
> Noise
> Radiation/electrical/radio/EMC
> Amenity
> Nuisance
> Product use

List of health and safety regulations
> Safety statement
> Signs
> Workplace
> VDUs
> Handling of loads
> Equipment
> Hazardous materials (exposure to)
> Pregnant/breastfeeding women

Clean air
Radiation
Carcinogens
Protective equipment
Noise
Dangerous chemical handling and storage
Electrical safety
Activities in transport
Hours of work
Activities in extractive industries

Outline of known steps to certification
Planning meeting
Assign initial tasks
Make policy commitment
Carry out preparatory review
Plan organization and personnel
Prepare Register of Regulations – both environment and health and safety
Carry out detailed examination of all operations and processes, including procurement for the production of the Environmental Effects Register
Draw up Environmental Effects Register
Design Environment Management System (Programme)
Design Environment Management Manual and all related documentation
Records, audits and reviews
Approach certification agency

12.2 Personnel involved

All the above has assumed that one person has produced the material shown – that is the headings shown above, not the actual work to be done. This is either the 'management representative' named in the standard or an outside consultant. As the management representative may not yet exist, it could be a project officer appointed to look at the standard, or even a person using his or her own initiative to approach management with the idea. It could be the quality manager looking beyond ISO 9000; it could be the human resources manager looking for a system to implement and control the health and safety procedures; it could be an engineer needing to implement the environmental directives. It could indeed be the company lawyers, but at a more general level.

It is suggested that the initiator, whoever he or she may be, presents the above outline at the first planning meeting, and obtains authority at that meeting to set up the project team, assign tasks, and so on. At that meeting also, senior management, preferably including the chief executive, should be asked to give the go-ahead. The policy statement can then be written, as of that date, ensuring that commitment has been formalized.

The first people at the planning meeting could be a consultant or the originator of the idea, the chief executive, the quality manager, the human resources manager and a senior engineer. These people could form the original project team, with one continuing as the project leader and some of the others assuming steering committee status as more workers are added to the project.

12.3 Identifying the first tasks

The very first task has been to devise the initial plan and appoint personnel. The next tasks are:

- Begin constructing the Register of Regulations. (Chapter 15 gives further details and a sample Register of Regulations is reproduced in Appendix 2.)
- Carry out the Preparatory Review.
- Produce an Issue Identification Summary (a very useful device for getting started). This is a macro-overview to ensure that no major issue is overlooked. (A sample is reproduced in Appendix 1.)

12.4 The Preparatory Review

What follows is an outline plan or format for a Preparatory Review.

1. Establish the following, with relevance to one's own company:
   - National environmental regulations.

- National health and safety regulations.
- Relevant international directives or standards (EC/ISO).

2. Evaluate significant environmental effects of activities/processes:
   - Product
     Procurement
     Packaging
     Utilization by end user
     Disposal.
   - Operations
     Draw up list of environmental interactions
     Draw up list of health and safety interactions.
   - Potential emergencies
     Project possible emergencies and consequences.
3. Evaluate existing environmental practices and controls
   - Vendor standards.
   - Product(s).
   - Operations.

12.5 Relevance of data to date

The Preparatory Review is a very useful exercise for the project team, which probably still comprises the originators of the project, to find out how much work is involved. There will be huge differences, for example, in what teams will be faced with in a software or publishing company compared with a food or chemicals manufacturer or a mining company. After the Preparatory Review the team will begin to understand the size of the project, how many insiders and outsiders are needed, whether to use specialist services for hazard or safety audits and so on. Depending on the company, and, in particular, on the size and scope of its activities, such as distribution, there will be considerable differences between the magnitude of the jobs and the scope of the issues involved.

12.6 Useful working documents

Two simple working documents are useful at this stage. These will be demonstrated again in the next chapter on implementing the Environment

Management System. These are:

> Issue Identification Checklist – Environment
> Issue Identification Checklist – Health and Safety

(There are examples of each in the next chapter.) These will form the basis of the Register of Regulations, and are, in turn, created from the useful lists in Section 12.1 above. They can also be used after the Preparatory Review, which may change them as some issues evaporate and others emerge, to design the final control documents.

Before designing the full checklists shown in the next chapter, one may find it useful to employ a type of early scoping study document to identify the issues at the broadest level, such as the following example.

Preliminary Issue Identification Checklist – Health and Safety

Name of regulation	Relevance to organization (some or significant)	Application area
Safety Health and Welfare at work Act 1992	Significant	Whole company – training – other see below
Safety/health signs	Significant	Whole company – installation – training
Handling of loads	Significant	Goods inwards Stores Shopfloor Dispatch

(And so on for all the issues as shown in the following chapter)

12.7 Plan for the installation of BS 7750

An outline plan for the implementation of a system to meet the requirements of BS 7750 may be drawn up as follows.

**Plan for implementation of BS 7750 system at
ABZ Corporation**

Submitted by:
John Doe
Consultant
10 September 1994

(Each of these sections would be on separate sheets)

List of contents
1. The project plan.
2. Outline of known steps from now to registration.
3. The Preparatory Review.
4. Environment Management Programme.
5. Approximate schedule.

1. The project plan

Task	Assigned to	Date complete
1. Draw up plan	BR	1 week
2. Identify first tasks	BR, PD	2 weeks
3. Assign first tasks	BR, PD	2 weeks
4. Start Preparatory Review – outline attached (After Preparatory Review, which should take about 1 month the manpower needs and approximate costs should be known)	BR, PD	immediate
5. Register of Regulations – inc. health and safety	BR, PD	1 month
6. Environmental Effects Register		3 months
7. Environment Management Manual An outline generic manual customized to the company as much as possible at this stage, i.e. including the processes relevant to the standard		3 months
8. Design and implement Environment Management Programme		12 months

9. Pre-registration audit
10. Application for certification   12 months

2. Outline of known steps from now to registration

These will become more complete as the project proceeds, but here for the moment are the known main steps.

> Planning meetings.
> Assign initial tasks.
> Preparatory environmental review – see attached.
> Write policy statement.
> Plan organization and personnel.
> Prepare Register of Regulations – environment and health and safety.
> Detailed examination of all operations and processes, including procurement for production of Effects Register. See lists below for guidance.
> Draw up Environmental Effects Register.
> Design Environment Management Programme.
> Environmental Management Manual and documentation.
> Records, audits and reviews.
> Approach certification agency.

3. The Preparatory Review

Establish the following:

1. The legislative and other requirements:
   - National regulations
   - EC directives
   - Health and safety regulations.
2. Evaluate significant environmental effects:
   - Product
   - Product ultimate disposal
   - Service procedures
   - Safety
   - Ergonomics of operations
   - Plans and procedures for emergencies
   - Procurement
   - Packaging
   - Staff health and safety

- Public health and safety
- Disposal of waste
- Litter
- Landscaping
- Emissions and discharges for plants, out-stations, head office, storage, transport and distribution
- Distributor's operations.

After the significant environmental effects are established – that is we have found what the significant issues are, it will be possible to complete the next section and establish approximate time scales and costs of the whole project.

This will mark the end of the Preparatory Review. The Review should look at a number of typical elements, such as the head office and depots. This stage is, in effect, a scoping study. The scope of the evaluation will embrace:

- Vendor standards
- Product
- Operations.

During the Preparatory Review the following list of possible environmental interactions will be used. This may expand as the scope becomes apparent.

Emissions
Effluent discharges
Solid waste
Toxic waste
Noise
Radiation/electrical/radio/EMC
Planning EIA demands
Amenity
Nuisance
Product use
Packaging
Public safety

The staff situation will also be considered under the Health and Safety Regulations, listed as follows:

Signs
VDUs

Lifting
Hazardous materials
Clean air
Training
Radiation
Protective equipment
Noise
Dangerous chemicals handling and storage
Electrical safety
Pregnant workers
Workplace
Equipment

4. Environment Management Programme

This will include:

All procedures identified, and controls specified
Full documentation
Implementation on the ground
Vendor assessment of supplier conformance to any directive demanding attention
Design considerations
Housekeeping
Inspection and testing procedures
Design of audits
Organization changes
Training

In summary, the integration of the necessary controls with the Register of Regulations.

5. Approximate schedule

After the Preparatory Review one will be in a better position to project dates and costs, but at this stage the project could demand a one year time span, that is, over a year as distinct from man weeks, because it takes time to obtain up to date regulations, to design and produce the necessary documentation, to train staff and implement procedures.

13

Implementing the Environmental Management System

The implementation of the Environmental Management System is our main purpose, but, as we have already seen, much other work needs to have taken place behind the scenes to prepare for this point. If, however, one has already implemented the standard in a sister company and has secured management commitment to repeat this in a new company, this chapter would describe the actual practical steps needed for implementation.

13.1   Review of steps already covered

Before implementation actually begins, three steps, which have been covered in previous chapters, are assumed to have been taken.

Management have already made the commitment and a policy statement has been written to record this fact in minutes of the meeting or as a paragraph for insertion into the Environment Management Manual. Resources have been allocated and at least the first members of the project team named. The project team has devised the initial plans and has completed the Preparatory Review.

13.2   Design and change processes

It is very important that companies engaged in the design of products and processes as well as manufacturing and services involve the design engineers at this stage. Problems encountered at the issue and control stages to be looked at shortly could well have an impact at the design stage. Designers cannot continue to do their work without a knowledge of the controls which the standard will demand; indeed, the very regulations may inhibit or ban the use of certain materials or processes.

In addition to this necessary interactive process between design and production, the standard demands control of the design and change process along the lines of ISO 9000. Any design company embarking upon the environment standard which has already implemented an ISO 9000 design and change control mechanism can simply use that mechanism to satisfy the demands of the new standard in this respect.

13.3   Setting the targets

Targets are the heart of the matter. The standard deals at some length with this in Annex A, under the heading 'Environmental effects'. It lists legislative and regulatory requirements and codes of practice. It introduces the subject of the 'environmental probity' of suppliers. It also looks for consideration of both normal and abnormal, or emergency, situations.

The standard is clearly asking for more than the meeting of the fundamental statutory requirements laid down in the Register of Regulations. In its earlier Section 4.5 'Environmental objectives and targets', it says: 'In addition to compliance with all legislative and regulatory requirements, other objectives and targets shall be identified after consideration of the environmental effects register' (which will be considered in Chapter 16) and it goes on to qualify this with financial and other requirements of the organization, 'in conjunction with the views of interested parties'. And, as if all this were not enough, it asks for continual improvement.

The revolutionary and far seeing nature of this standard is revealed in this paragraph. What is required is that an organization uses the regulations as a minimum objective and then builds upon these both with a policy to achieve something better and a willingness to listen to the requests of interested parties, presumably the community and perhaps also customers. Also, adopting a policy of continuous improvement sets the organization on a course of becoming cleaner and cleaner.

The introduction of codes of practice recognizes that there will also be existing and evolving industry standards for specific industry groupings. The commitment of the organization to the specification of numerical targets for matters such as pollutants to air, land or water, and waste reduction is also asked for.

In dealing with emergency situations, considerations should also be given to the potential effects of fire, traffic accident, explosion, flood or malicious damage, so a company cannot simply walk away from these and blame acts of God or terrorism.

13.4 Issue identification

The standard asks at this stage for issues, or environmental effects, to be evaluated in the light of environmental policy. However it may not be practical at this early stage of implementing the environment standard to go beyond regulatory requirements, unless the organization is already operating to high environmental standards.

We need the Register of Regulations to define the effects of operations and to construct a Register of Environmental Effects, but as we identify effects we may realize the implications of other regulations not appearing, hitherto, to have significance for our operations. We begin, therefore, with the construction of the Register of Regulations, derived in the first instance from collecting all of the apparently appropriate regulations, as described in Chapter 15 and the Preparatory Review.

At this stage the Issue Identification Checklist for both environment and health and safety can again be employed. It was used initially in the Preparatory Review; now it can be filled out more thoroughly. To oversimplify the process for the sake of illustration, one would take the Register of Regulations in one hand and Issue Identification Checklist in the other, walk around the organization and by scrutinizing the activities and processes in the light of the regulations, identify the issues. The complexity of this will vary greatly from industry to industry, as can be seen from the checklists.

Figure 13.1 may help to put the matter of issue identification into place.

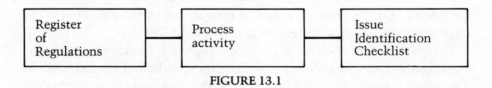

FIGURE 13.1

Sample Issue Identification Checklists follow, firstly for environment (Figures 13.2 and 13.3), and secondly, a sample Issue Identification Checklist for the health and safety issues (Figure 13.4).

Effluent discharge checklist (Environment)

Initials:_____ Date:_____ For000
Page 1 of 1

Issue	Action	Pass	Fail
Particulates			
BOD			
SS			
COD			
OILS			
MBAs			
SO_4			
Cl			
NH_3			
NO_3			
PO			
Metals			

FIGURE 13.2

113

Emissions checklist (Environment)

Initials:_____   Date:_____   For000
Page 1 of 1

Issue	Action	Pass	Fail
Particulates			
SOx			
NOx			
HCl			
CO			

Other checklists

EC Eco-scheme
Physical planning
Waste
Packaging
Visual display unit
Software ergonomics
Nuisance and noise
Trees, amenities, landscape,
and wildlife
Use of materials
Use of energy
Product life cycle

FIGURE 13.3

Envirocompany - Work place checklist (Health and safety)

Initials: _____   Date: _____   For000
Page 1 of 1

Issue	Action	Pass	Fail
Ventilation	Check for adequacy consider-volume of workplace and number of occupants		
	Check for air contamination from processes occurring in the work place		
	Check for drafts		
Temperature	Check temperature seated work not below 16°C		
	Check temperature physical work not below 13°C		
Lighting	Check for sufficient natural and artificial lighting		
	Check for sufficient emergency lighting		

Other checklists

Work equipment
Visual display units
Manual handling
Personal protective equipment
Pregnant workers
Signs

FIGURE 13.4

115

13.5 Procurement

Procurement is one of the most interesting elements in the standard. Perhaps what is not said about it is more interesting than what is said, so much is simply suggested or implied. Under Section 4.8.2, Control, procedures are asked for to control the activities of suppliers and sub-contractors to ensure that their services and products comply with company policy. The Annex makes this more specific, first mentioning the 'environmental probity' of suppliers and suggesting that, where it is not possible to simplify 'detailed registers', alternative sources of supply can be compared.

Under Operational Control in the Annex, it becomes more specific about procurement activities, firstly admitting that suppliers may not always be able or willing to provide the information needed. This appears to be a little less confident than the rest of the standard, particularly as the authors must be well aware of how little choice suppliers had in dealing with the demands of ISO 9000. All it takes is one important customer implementing the standard to impose requirements on suppliers and the epidemic of the standard is under way.

The demands on procurement here are very similar to those required by ISO 9000. Measurements must be identified as well as the required specified levels of delivery. The measurements must be specified with regard to time and place, and documentation is demanded. Indeed, the whole ISO 9000 procedure for procurement, or for other assessment systems, could be used as it stands by simply substituting environmental specifications for quality specifications. Packaging materials are perhaps the best example, but demands could just as easily apply to how one's electricity is produced, for example is the power utility scrubbing its emissions from its coal-fired station or is it contravening international or national regulations? The standard also expects organizations to examine indirect effects such as those of extractive industries supplying raw materials and even the activities of companies in which an organization invests capital.

13.6 Audits

This is one of the areas where the ISO 9000 standard can be used, with no need for the creation of a separate auditing system. The main reason for this is that the audits for both standards are based on the ISO 10011 series of standards which give guidelines for auditing management systems. Environmental Management Audits are required by the standard after implementation as part of the continuing operational Environment Management System. At present we are still discussing the implementation project, so the audits discussed here are those which may be useful at the project stage.

The one audit specified by the standard for this phase is the Preparatory Environmental Review (see Chapter 12). This is very close to the kind of preliminary audit carried out by companies before embarking upon the ISO 9000 standard. Other kinds of assessment, however, may be found necessary with the environment standard. Three areas of activity can be cited as examples.

Firstly, one may need to carry out a risk assessment to establish what emergency possibilities exist and to plan for them accordingly. This may involve using outside expert consultants. Secondly, the actual status of the safety of the workplace for staff, which now features in health and safety legislation, may require auditing by independent expert safety consultants. Thirdly, a service organization or a widely diversified company with many branches or outlets may need a large scale study of its operations to address all aspects of operations.

Once the standard is in place, the regular audits and reviews come into play. For further information on this subject, consult *ISO 9000*, published by Gower.

13.7 Operational controls

The operational controls are those parts of the Environment Management System which allow it to be measured and managed. These will consist of full operational procedures applying to each control function, for example, the control of emissions, the disposal of scrap. These are all, in turn, shown as master copies in the Environment Management Manual, or filed in separate procedural manuals, cross-referenced to the Environment Management Manual. The procedures must also cover the relevant activities of both suppliers and sub-contractors.

In all cases of automated and process plants, the procedures also apply, whether computerized or automated themselves. The actual criteria needed for meeting regulations and policy targets must also be stipulated in the procedures. The equipment used to maintain the controls must also be subject to a system of calibration and measurement. The ISO 9000 procedure for doing this in the quality management system can be used as it stands in this regard or simply expanded to include the environment measurement equipment. All laboratories used in this process must also be fully controlled along ISO 9000 lines. Indeed, the whole ISO 9000 documentation system could embrace the controls needed and *the control of those controls* for the Environment Management System.

Similarly, the procedures in ISO 9000 for dealing with non-compliance and corrective action can be extended to cover the environment standard. Mechanisms must be installed to authorize action in the case of non-compliance, to restore compliance and prevent recurrence, and a system must exist to measure the effectiveness of these mechanisms. Sample documents for dealing with operational controls (Figures 13.5 and 13.6) and element monitoring (Figures 13.7 and 13.8) follow.

Envirocompany - Effluent discharge control (Environment)

Initials:_____   Date:_____   For000
                                                            Page 1 of 1

Issue	Procedure *	Parts/unit
Particulates		
BOD		
SS		
COD		
OILS		
MBAs		
SO_4		
Cl		
NH_3		
NO_3		
PO		
Metals		

* See procedures manual
  or environment manual

FIGURE 13.5 Issue control mechanism.

Envirocompany - Emissions control (Environment)

Initials:_____   Date:_____   For000
Page 1 of 1

Issue	Procedure *	Part /Unit
Particulates		
SOx		
NOx		
HCl		
CO		

* See procedure manual
  or environment manual

FIGURE 13.6 Issue control mechanism.

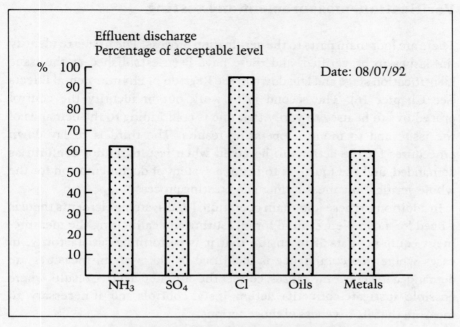

FIGURE 13.7 Element monitoring: weekly.

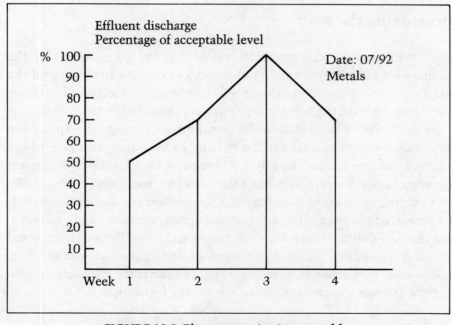

FIGURE 13.8 Element monitoring: monthly.

13.8 Verification, measurement and testing

There are four main parts to the verification process. The first is to identify the issues to be verified, and these have been established at the issue identification stage and laid down in the Register of Environmental Effects (see Chapter 16). The second is to work out or identify the control procedures to be used to verify that one is conforming to the demands of the issue and to monitor operating results. The third is to lay down procedures for the actions to be taken when results do not conform as demanded, and the fourth is to instal a system of documentation for the whole verification, measurement and testing process.

In addition to these four main operational and systems elements there is a need for a complete system for measuring and calibrating the measurement equipment itself to ensure that it is performing satisfactorily. In cases where non-compliance is identified by the system, measures are demanded to determine causes, correct the current defective results where possible, instigate corrective action, instal controls and, if necessary, to measure the effectiveness of those actions.

13.9 Organizing the staff

The 'management representative' called for by the standard is the environment manager or quality manager, or both. As already noted the signs are that the quality manager will handle both quality and environment matters with the latter being considered as a quality issue also.

As in the ISO 9000 standard, the environment manager is expected to have sufficient knowledge for the job and to be able to delegate where necessary to 'nominated deputies'. There must be no conflict of interest between his or her environment role and other functions, for example, cost-cutting or production demands. The standard sets out possible roles for senior management, the environment manager, finance and personnel, and then goes on to name most other functional areas. Finance is expected to identify costs and benefits, personnel – communications and training, and senior management – policy. The environment manager is also expected to monitor environmental issues, and legislation.

13.10 Documentation

There are as we have seen three main manuals: the Register of Regulations, the Register of Environmental Effects and the Environment Manual. The third is the master containing references to the other two, the policy statements, the outline of the Environment Management System and is cross-referenced to the procedural manuals or individual procedures. It must also contain all the control documents, including the controls for the revisions, updates and amendments to the system and to the Environment Manual itself.

14

Obtaining Certification

This book was written after the 1992 publication date of BS 7750, but before the first formal certification scheme was in place for either BS 7750 or the EC Eco-Management scheme. What now follows is an assessment based on the current information available and on the author's knowledge of ISO 9000.

There is an opportunity here, however, for early achievers of the standard, for there is nothing to stop them declaring that they are operating to the standard once the systems it asks for are installed.

14.1 Certification and registration

This complex subject is made more difficult by the fact that it is changing at the time of writing as attempts are being made to harmonize practices within the new Single Market. Within the EC, the harmonization of testing and certification services is a natural follow-on to the process of harmonizing standards. Where a single standard, or set of standards is agreed, this can have practical consequences only if common certification schemes exist, and if each national scheme is accepted throughout Europe or internationally. A number of US and other overseas agencies are qualified to certify to the ISO 9000 standard, so we can assume that any country whose agencies can certify to ISO 9000 has the infrastructure to set up certification schemes to certify to an environment standard. What is far from clear is whether departments of industry or environmental protection agencies will supervise the schemes, although the latter seem more likely to do so.

The main purpose of harmonized certification is to allow products free access to the markets covered by the harmonization, so that repeat testing is not required in the countrles of destination. The main mechanisms in Europe are legislation through EC directives, which become national law in each country, supported by harmonized product and management standards. The CE Mark, for example, placed upon a product, denotes that the product conforms to the requirements of all of the applicable EC directives and that testing has taken place in accordance with rules laid down in the directives.

The CE mark denotes that the product conforms to essential requirements and is a 'passport' to European markets. It is not, however, an indication of quality, but may in the future demand conformance to an environmental standard.

To help meet the demands of the directives one uses European Standards (ENs) which are also published as national standards within each country. Each national government establishes the testing and certification systems and notifies the EC Commission of the names. These are known as 'Notified Bodies'.

14.2 The certifiers

The so-called 'Notified Bodies' can be certification agencies, test laboratories, and inspection agencies, usually appointed by government, although this is not yet clear. They are required to be totally independent and must conform to another set of European Standards known as the EN 45000 series, which have been put in place to provide a uniform system of accreditation across the EC.

The committees responsible for the harmonization and standards writing processes have had to guard against 'groups of fanatics', as they are described by some, trying to impose unrealistic expectations, and this will be needed increasingly because of the ease with which those advocating simplistic solutions have been able to take over environmental groups and bring them into disrepute. The advocates of standards need to represent a large constituency, which should include the market. They need to be practical and realistic and to provide a continuity in their participation.

Manufacturers inside and outside Europe need to be aware that CEN, the European Committee for Standardization, is constantly at work devising new standards which become ENs. Once the EN is realized, all CEN states must comply with its requirements, and any outsider trying to sell into the relevant market without knowledge of the EN may become excluded, and will be excluded in the case of public procurement.

If, for example, local authorities are buying and the product sought is covered by a new EN, prospective suppliers must ensure that their offerings comply. This could require re-design, re-tooling and re-manufacture. If one is a European subsidiary or licensee of a US or Japanese company, this means both acquiring a knowledge of the evolving ENs and passing the information back to the principal.

This is the case for specific EN product standards, but for a management standard such as ISO 9000 or BS 7750, one may also find that codes of practice, however voluntary, are also expected to be adopted. In the case of BS 7750, a very good example is ISO 9241, the software ergonomics standard, becoming a desirable code of practice for software developers attempting to acquire either ISO 9000 or BS 7750.

Even the Construction Products Directive which was implemented in 1991, and which applies to all products to be used in a permanent manner in building or civil engineering works, now lays down environmental requirements, which may shortly, in turn, demand control by a management standard.

14.3   The current situation

In its brochure advertising BS 7750, BSI states: 'It may help companies meet the criteria of the proposed EC regulation'. It goes on: 'BS 7750 is consistent with the draft European Community regulation to set up a voluntary scheme on environmental auditing ("eco-auditing"). Use of the standard may be seen as a first step to meeting its requirements.' This rather modest statement would lead one to believe that the draft Eco-

Management scheme demands more than BS 7750: however, a reading of both does not bear this out, nor does the first BSI sentence quoted above.

In a separate paragraph in the brochure it reveals that ISO and CEN are working on the development of an international standard on environmental management systems. It is clear from this that BSI expects, or hopes, that BS 7750 will follow the success of BS 5750, and result in an ISO and EN standard.

The press release announcing the launch ventures into the area of accreditation saying, 'It does not specify expected levels of organizational performance, but rather specifies a standardized management system that is capable of independent assessment and verification. Accreditation of any certification scheme linked to the standard will be controlled by the Department of Trade and Industry (DTI), who have indicated that they will not accredit any scheme until the position of European initiatives is much clearer.'

This is an extremely significant statement as it places the UK's DTI, not the DOE, in the role of certification watchdog, as if it were saying that British industry and its exports need BS 7750 and this is the business of the DTI. Whether this position continues remains to be seen, but it is virtually certain that departments of the environment and especially EPAs (environmental protection agencies) elsewhere will want to administer this standard and how they do so may determine the success of whole economies.

The BSI newsletter accompanying the launch carried a joint statement from the DTI and DOE welcoming the standard and stating that it was essential that it be compatible with the EC Eco-Management regulation. It added that the arrangements for accrediting the companies which offer certification to the standard would not be put in place until the European position was clear and 'the government is in a position to take decisions on the accreditation arrangements which will be required under the EC Eco-Management regulation.'

Meanwhile, two other groups had moves under way to get involved in certifying the certifiers. These were the Association of Certification Bodies (ACB) and the National Accreditation Council for Certification Bodies (NACCB). These, however, were trying to become the overseeing accreditation body for the UK, but in an open single market certified assessors and certification agencies can operate anywhere.

At the time of writing, several private certification companies were offering to certify to BS 7750. Some of these were offering their own symbols to companies, which they had certified to the standard, and were expecting that these would receive retrospective accreditation when the formal scheme came into being.

This is a good example of the two main aspects of a management standard: firstly, its adoption and implementation by a company, and, secondly, independent certification, in this case by a company prepared to offer its own symbol of conformance pending the official conformance scheme.

14.4 The process

The certification process is, or should be, a copy of the ISO 9000 process. Some certification agencies ask to see a copy of one's ISO 9000 Quality Manual before agreeing to send inspectors. It is not clear if this practice will continue with the environment standard and whether they will look for the Environment Management Manual or the Register of Regulations. The latter may be quite a heavy volume.

The procedure should be to implement the system fully, carry out a pre-certification audit, using a checklist such as the one displayed in Figure 14.1 and then apply for inspection and certification.

Envirocompany - Pre-registration audit checklist	For000

Initials:_____   Date:_____   For000
Page 1 of 1

Element	Action	Pass	Fail
Policy,management and document-ation	• Is there a clearly defined environment policy? • Is there a management system for executing the policy? • Is the management system clearly documented? • Is there a document control system? • Have responsibilities and authority been stated? • Is there a training and education program? • Are there clearly defined procedures? • Is there a register of issues? • Is there an environment management manual? • Is there a register of regulations? • Are the results of monitoring being recorded and presented to management?		

Issue:
    Waste
 	
    Toxic waste
 	
    Emissions
 	
    Effluent discharge
 	
    Environmental impact
 	
    Noise and nuisance
 	
    Software ergonomics
 	
    Trees, amenities, land-scape, and wildlife
 	
    Physical planning
 	
    Energy
 	
    Materials
 	
    Product quality
 	
    And one for each other relevant issue
	• Check that there is a procedure for all relevant issues either in the Environment Manual or separate procedures manuals		

FIGURE 14.1 Pre-certification (or pre-registration) checklist.

Envirocompany - Pre-registration audit checklist   For000

Initials:_____   Date:_____   For000
                                                                    Page 2 of 2

Issue	Action	Pass	Fail
First year only Was Preparatory Review done? If so is report available for inspector? **Every year** Is there a health and safety system in place? Is there a plan for emergencies? Is there an energy conservation programme? Has an internal audit been carried out? Is there a staff training programme?			

FIGURE 14.1 (continued)

131

PART IV

Implications of the Standard

15

Register of
Regulations

This is the record of all of the relevant regulations, policies and targets for
meeting environmental standards. It can simply be a ring binder containing both the actual regulations where relevant and statements about each
of the 'issues'. The issues as we shall see in the Register of Environmental
Effects are those environmental issues that relate to the organization.

15.1 Localization of regulations

This is an important point for readers outside the European Community.
EC directives form the basis for most of the national legislation in the EC
member states. A directive is published and then followed by a statutory
instrument, which may also be called a ministerial regulation or act of
parliament, making the EC directive law in the national state.

This book is also intended for readers outside the EC, but the EC
directives are used as they are the first co-ordinated 'harmonized' body of
agreed rules which are likely to influence international regulations. It is

also this writer's belief that, as with ISO 9000, Europe is leading the way in the process of the internationalization of agreed responses to environmental demands.

At the same time, however, the non-EC reader should treat the EC examples as examples and no more; even the EC reader may find that the directives given have changed by the time this is read. All readers should accept that adopting the standard means sourcing the local regulations and establishing a register. The examples given here are, for the most part, fairly generic. Every country will have issues such as planning, emissions, effluent, noise, amenity and health and safety. Some may have extra issues not covered here. The onus is on all project leaders implementing the standard to identify and record the relevant regulations in their own territories, including member states of the EC.

15.2 Policy and targets

A second important point is that the standard is asking for more than the meeting of regulations. The regulations may be pending or still not on the drawing board, but there may be industry codes of practice. Some standards will be covered by legislation, others not. It will be found, for example, that even the voluntary ISO 9000 standard becomes an environmental issue in cases where the product has an environmental implication, for example, within a packaging company, and requires ISO 9000 as the guarantee of the maintenance of product standards. The Register of Regulations can contain legal regulations, standards, codes of practice, and policy statements. What matters is that each major environmental issue be dealt with, and this chapter and the sample register, reproduced in Appendix 2, attempt to list most of the issues.

15.3 Sourcing the information

One can start with the samples shown in this chapter and then make ready a large and sturdy ring binder. At this stage the actual collection of the regulations begins. It may not be necessary to collect and file all the actual regulations which will apply for a period of time. For example, local authority planning and building regulations will apply to all construction work, including extensions, but it is sufficient for the register to refer to these by name and number with a commitment that they will be adhered to whenever any building work is embarked upon. Similarly, some regulations, such as waste removal, are in the hands of the local authorities: it is sufficient for the register to state that only the local authority shall remove waste.

So what are the sources? All EC member states will have an accessible EC information office for locating the directives on environment, health and safety. Each will also have a government publications office. Most other countries will have the latter and many, including the EC, will have both an environment information service and a health and safety authority. In both of the last cases the enquirer may be fortunate also to obtain overviews which list all of the directives and regulations. These overviews can be inserted into the register with the relevant issues highlighted. They demonstrate that the company has thoroughly examined all of the potential issues.

15.4 Making the information available

Making information available is a rather more difficult matter. The standard asks for an open approach to interested parties and a communications system. The usual demonstrations of conformity to the trinity of quality, environment and health and safety are declarations of policy on the walls of the organization's lobby, and the accompanying display of certificates of registration to ISO 9000 and BS 7750. It is a bit more difficult to make the Register of Regulations available. Perhaps a video on show in a visitors' room would be a practical way of meeting this demand. One irony is that many companies who have made information available to the community find that no one ever asks to see it.

15.5 Sample Register of Regulations

A sample Register of Regulations is reproduced in Appendix 2, p. 169.

16

Register of Environmental Effects

This is where the general, probable and likely issues established in the Register of Regulations are defined in a manner so specific that controls can be created. The process, however, is not as a result of a direct transfer between the registers, but as a result of an issue identification process beginning with or before the preparatory review.

16.1   Overview of the system

Figure 16.J shows diagrammatically where the Register of Environmental Effects fits into the system and a sample Register is reproduced in Appendix 3.

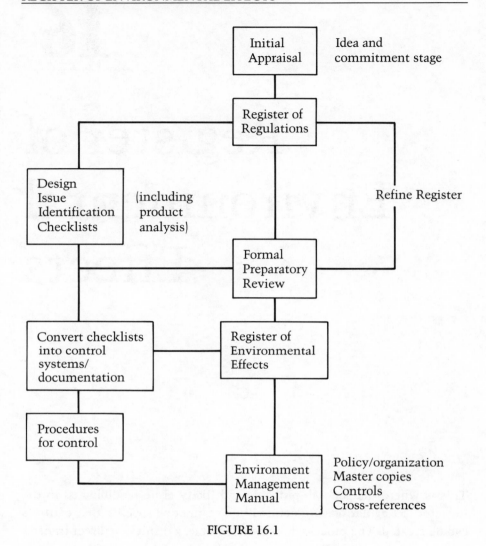

FIGURE 16.1

In theory, the Preparatory Review would appear to come before the Register of Regulations. It will be more practical to build the Register first by simply incorporating all of the obvious legislation. One will know whether air, effluent or wastes are involved, for example, and obtain the relevant acts and regulations.

The checklists evolve from early Issue Identification Checklists used for the once-only Preparatory Review, and filed for inspection, and then converted into control mechanisms, which may be hard copy or programmed controls. If the latter, hard copy reports or other output, perhaps in graphical form, need to be maintained for inspection.

Checklists developed during the Preparatory Review are used to establish exactly what the effects of each particular relevant issue are within the site.

16.2 From issue identification to effects

An Issue Identification Checklist for incinerator emissions is demonstrated in Figure 16.2.

The Preparatory Review using expert tests established that we have the issues illustrated in the figure.

Envirocompany - Emissions checklist (Environment)

For000
Initials:_____   Date:_____   Page 1 of 1

Issue	Action	Pass	Fail
Particulates			
SOx			
NOx			
HCl			
CO			

```
        Other checklists

EC Eco-scheme
Physical planning
Waste
Packaging
Visual display unit
Software ergonomics
Nuisance and noise
Trees, amenities, landscape,
and wildlife
Use of materials
Use of energy
Product life cycle
Effluent
```

FIGURE 16.2

141

16.3 The controls

How do we control these issues? We implement whatever measures are necessary, manual or automated, and then we set up both a measurement and control system, showing targets or licensed discharges against results.

Issue – incinerator emissions

One checklist for each output, e.g., Particulates

Actual emission	*Licensed amount*
Particulate	Our actual emission
SO_x	situation for each
NO_x	output
HCl	
CO	

And, on a daily, weekly or monthly basis, we maintain documented results in graphic or other usable format.

The elements of the environment management control system are the use of control documents, such as the one shown above in Figure 16.2, a monitoring log, as shown in Figure 16.3, and the issue checklists from which the control documents are devised.

The overall controls are as follows:

- Use the monitoring log to make the necessary checks.
- The monitoring log specifies the relevant checks and the frequency of the checks.
- Enter the initials of the checker under the appropriate date when the checks have been performed.

A monitoring log summary is shown in Figure 16.3.

Figures 16.4 and 16.5 show the documented results of emission checks in graph format.

Envirocompany - Monitor log summary (Environment) For000

Name: _____   Date: _____   Page 1 of 1

Issue	Form	Date											
		4/1	4/2	4/3	4/4	6/5	4/6	4/7	5/8	4/9	4/10	4/11	4/12
Physical planning Waste													
Packaging													
VDU													
Software ergonomics													
Nuisance and noise													
Trees, amenities, landscape, and wildlife													
Emissions													
Effluent discharge													
Use of materials													
Use of energy													
Product quality													

FIGURE 16.3

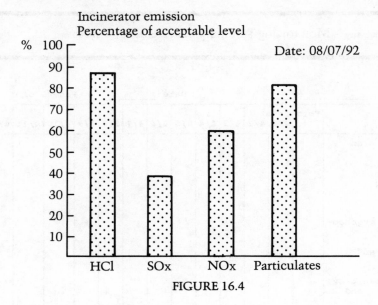

FIGURE 16.4

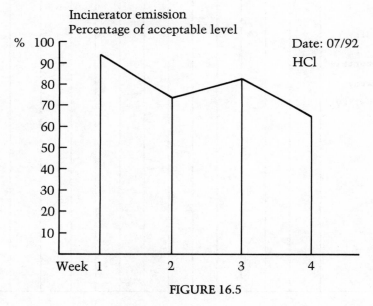

FIGURE 16.5

16.4   Sample Register of Environmental Effects

A sample Register of Environmental Effects is reproduced in Appendix 3, p. 191.

17

Product Life Cycle

At first glance it may not seem like an environmental issue, but the product itself may be the biggest environmental issue of all, not just in the issues it gives rise to in its production, but also in its use with end-users and in its ultimate disposal.

17.1 The product issue

The product, and the service also, may be a main environmental issue. A good example is the production of mercury soap, which is banned for use in many countries and the manufacture of which is banned in Europe. Mercury soap has a reputation in certain African countries for making dark skin lighter, but its use can cause mercury poisoning. One could have a mercury soap factory built to very high standards, with no damaging environmental emissions or discharges or toxic waste, and in which staff work in a safe and healthy manner. Such a plant could even have its quality management systems registered to ISO 9000; and at the same time the product could be marketed in an unethical manner, as a cosmetic, and be dangerous to the end-user. For the manufacturer this is an environmental issue – a dangerous product or a product being used which endangers people.

This is an extreme, but unfortunately not exaggerated, example of an environmentally undesirable product, in its product-use sense, as used by people, as distinct from undesirable effects on the environment, as with the products used in refrigerators and as propellants. Other less extreme examples would be found in those products which present ultimate disposal difficulties, which can be dangerous to health, if only through misuse, or to children, or where the packaging is undesirable.

17.2   Product analyses

Either at the Preparatory Review stage, or at an earlier issue identification stage, the product must be analyzed, that is actually taken to pieces, at least on paper, and probably physically as well, to identify each of its separate sub-components. It may be convenient to do this for both the product and its packaging, especially where different packaging accompanies different products in a range. What follows is such an analysis for a well-known and apparently simple product – a software package, which is made up of magnetic disks, paper and cardboard wrapping, and an instruction manual.

Product

Software Package (Make up)

Box
   Cardboard type 1 (two walled corrugated)
   Cardboard type 2 (single walled)
   Ink
   Glue

Pressure sensitive labels:
   Paper
   Plastics
   Ink
   Glue

Contents
 Paper
 Cardboard type 2
 Ink
 Glue
 Alloy Staple

Floppy disks
 Plastics
 Cloth
 Alloy
 Metal oxide

And here are examples of product analysis checklists:

Analysis (process)

Product analysis (end use). A sample checklist is shown in Figure 17.1.

Envirocompany - Product analysis end use (Environment)

Initials:_____   Date:_____   For001
                                                    Page 1 of 1

Product			
Component	Environment issue	Regulation	Standard

FIGURE 17.1

17.3 Relationship with product standards

It can be seen from the foregoing documents that both regulations and standards are 'components' demanding compliance within any management standard. One cannot claim either an overall quality or environmental management standard unless individual components satisfy their own relevant standards. A concrete products company looking to achieve ISO 9000 would have to ensure that its products were manufactured to each of the specific concrete product standards – blocks, beams, precast units and so on. Similarly a company wanting accreditation to BS 7750 needs to ensure that all the components of its product conform to individual environmental or safety or health standards, such as the IEC or EC standards on the safety of electrical equipment, toxicity of paint in children's toys, flammability of components, carcinogen content, radiation and so on. These are all elements which have relevance in product use. These will now be looked at in relation to the health and safety regulations.

17.4 Relationship with health and safety

It may be instructive simply to consider the implications of one's product being a factor, in the context of its end use, in the EC Health and Safety Regulations. Take, for example, the workplace regulation which demands a safe and healthy work building. If one is manufacturing components or systems, such as ventilation or air-conditioning devices, which directly affect the conditions demanded by the Workplace Regulation, while they are health and safety issues for the occupiers of the building, they are environmental issues, affecting use of product, for the manufacturer.

The same applies to all of the equipment covered by the Work Equipment Regulation, for example, in relationship to protective equipment, electrical shock and so on. Carcinogens in materials or products and hazardous substances similarly become environmental issues for manufacturers, while remaining health and safety issues for users.

An excellent example is the EC VDU directive, which is a health and safety issue for users, and parts of which, such as the environment of the work-place and seating, are exclusively issues for users, but much of which is also the responsibility of both hardware and software manufacturers. And, if one wishes to question how 'compulsory' this is for such manufacturers, one simply has to consider that if one's product does not allow a user to comply with a compulsory health and safety regulation, even if one's contribution is only the software creating too many lines on a screen, that user must by law stop buying that product. It quickly becomes apparent that much of the so-called 'voluntary' nature of an environmental management standard, such as BS 7750, is, in fact, mandatory. The manufacturer of the product or component may not be prosecuted, but the user may. The product element of the standard, therefore, can be regarded as virtually mandatory, more so than has been the case hitherto with ISO 9000, where, apart from electrical and other specific safety standards, market pressures only apply. With ISO 9000 now, however, the health and safety regulations are demanding a similar compulsory compliance from manufacturers whose outputs affect them through their use in the marketplace.

17.5 Ergonomics

Ergonomics – the study of people in relationship to their working environment – has become more popular with the proliferation of information technology. A significant part of the VDU regulation deals with ergonomics, but the subject goes beyond the VDU and embraces software affecting VDU display.

There is a 1993 standard – ISO 9241 – the software ergonomics standard. This addresses those elements of software relating to the integrity and accuracy of its use. Consider software used in a critical situation, controlling a complex processing plant or flying a passenger aircraft. The accuracy and *usability* of the messages produced by the software are of critical importance in product use, with the potential for disaster, and, at the very least, for the kinds of accidents and emergencies demanding attention in BS 7750. Software is now ubiquitous in our daily lives, and is probably an outstanding example of the ergonomics demands of one product in an environmental context.

Thinking this through fully, ISO 9241 becomes a fundamental demand on all hardware and software products in systems attempting to adopt BS 7750. It is an emerging international standard, and therefore a code of good practice, and the standard asks for the adoption of codes of practice. The standard may lead to a compulsory ergonomics regulation. Meanwhile, the VDU related elements of it are already law. This may all come as a bit of a shock to software and hardware manufacturers who see only such issues as discharges, emissions, waste and noise as being environmental. When even the package on the product is an environmental issue, how much more so is the product itself.

The issue of the ergonomics of product usage and, in particular, the implications for software developers, if their products are used in potentially critical situations, is actually highlighted by BS 7750, by its application to staff, operational integrity and public safety. These are now very important issues for software developers.

Software is merely a good example of the relationship between ergonomics and environment. ISO 9241 and the VDU regulation are excellent examples of the implications of product use conveyed by a strict reading of BS 7750, or any international environmental code of practice, and they cast light on both product liability and on how one may use a standard to protect oneself against claims of negligence. But ergonomics go beyond VDUs and software. There is now a new keyboard standard, with an ISO number, and standards involving ergonomics apply to many other products.

One area which strikingly illustrates this is that of laser applications. EN 60825 is the EC Laser Safety Standard which has an equivalent standard in each member state – BS, IS and so on. This standard is based on an original 1984 standard published by the International Electrotechnical Committee (IEC 825) adapted by a 1990 amendment. At first glance this may appear to be a non-compulsory standard like thousands of others, but let us consider its implications.

Firstly, under the EC Health and Safety Framework Directive and its equivalent national acts, every employer is compelled 'to ensure, so far as is reasonably practicable, the safety, health, and welfare at work of all employees'. In addition there is an explicit requirement to provide 'such information, instruction, training and supervision as is necessary'. Thus, the Framework directive states the general compulsion on employers, but then the two specific regulations of safe workplace and safe workplace equipment require that all known hazards be guarded against. An existing laser safety standard therefore automatically becomes compulsory under the health and safety regulations.

From the design and manufacturing point of view, all laser equipment, and its software, now become environmental issues from the point of view of product use, requiring not only that the full requirements of EN 60825 are designed and manufactured into them, but also that they are accompanied by safety instructions produced by the manufacturer and inserted with the product. It is difficult to see how anything short of this could satisfy the requirements of BS 7750.

17.6 Compulsory standards

Compulsory standards are so named to distinguish them from the health and safety regulations. Compulsory standards are those applying to certain everyday products found in the marketplace, many intended for the home, which are supported by a government regulation, often called a 'Ministerial Regulation' or 'Statutory Instrument. Typical products covered by compulsory standards are children's nightdresses (for their flammability), electrical devices, babies' prams, pushchairs and dummies, children's anorak hood cords (for breaking point if caught in swings), furniture (for ignitability) and so on.

As with ISO 9000, it is not conceivable that a manufacturer could achieve a management standard for production and processing if the products being produced did not conform to existing compulsory standards.

17.7 Packaging

Directives and regulations for packaging are under way at the time of writing. The implementors of the standard need to check with local standards bodies at the time of their projects to ascertain the current position *vis-a-vis* packaging. At the date of publication of this book, the Eco-Labelling Directive had not yet emerged, but it needs to be addressed when it is published. Other packaging issues that need to be addressed are the avoidance of over-packaging, the use of packaging materials that do not harm the environment or reduce scarce resources and the recycling of packaging waste.

17.8   User environmental effects guides

An excellent idea for those manufacturers or service companies whose products have a consequence for the environment is to insert user guides into the packages which contain the products.

An example of a four-page Hewlett-Packard leaflet advising customers about the recycling of HP toner cartridges is reproduced in Figure 17.2, starting on the next page.

HEWLETT PACKARD

Help Clean the Environment

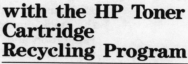
with the HP Toner Cartridge Recycling Program

When you return a used toner cartridge, you reduce landfill waste and help support the National Wildlife Federation and The Nature Conservancy.

National Wildlife Federation

FIGURE 17.2 Four-page leaflet *Help Clean the Environment*, reproduced with kind permission from Hewlett-Packard Company.

153

The increase of waste in our environment is a serious concern for all of us. With your help, Hewlett-Packard would like to contribute to the solution with a program for recycling used laser printer toner cartridges.

This program will serve several useful purposes. First, it will reduce the volume of plastics entering our landfills.

Second, it will conserve precious industrial resources, as the cartridges will be completely disassembled, with most of the parts either re-used or melted to be reused as raw material.

And third, it will help protect wildlife as well as the environment, because the National Wildlife Federation and The Nature Conservancy will share equally a $1-per-cartridge contribution from our cartridge vendor for each cartridge that is returned.

Becoming a part of this worthwhile program is very simple. When your cartridge is empty, just follow the simple instructions on the opposite page.

We appreciate your cooperation in contributing to this worthwhile program. A small effort on your part will result in one more step toward a cleaner environment.

Please take your step today.

The cartridges are not collected for refilling.

Defective cartridges under warranty should be returned to an authorized dealer or service facility as provided in the warranty. Do not send them by U.P.S. return label.

FIGURE 17.2 (continued)

U.S. Cartridge Return Instructions

Option 1: Single Cartridge Return Instructions

Step 1: Remove the existing used cartridge from your printer and slip it into the bag from your new cartridge. Seal the bag, then place it into the cartridge box with the foam end inserts. (The box and foam will be recycled as well.)

Step 2: Carefully seal the box with tape and apply the enclosed U.P.S. Authorized return label. Please cross out all old labels.

Step 3: Give the cartridge to any U.P.S. delivery driver. Your cartridge will be returned at no charge to you.

Option 2: Volume Return Instructions

This option is available for companies to return more than four cartridges at a time. We encourage you to use this approach if possible, since it's more efficient for shipping purposes.

Step 1: Place the individual cartridge boxes into a larger carton that meets the following U.P.S. criteria:

> 130 inches maximum in length and girth
> (L+2W+2H)
> Maximum length of 108 inches
> Maximum weight of 70 pounds

(We recommend a carton that holds 8 to 10 cartridges. If you can re-use an existing shipping box, please do!)

Step 3: Prepare the collection carton(s) for shipment. Tape each carton closed and remove any existing shipping labels. Apply the enclosed U.P.S. Authorized return label.

Step 4: Place the collection carton(s) in your receiving area for pickup by your U.P.S. delivery driver.

The U.P.S. label is available for U.S. returns only. Toner cartridge returns are not tax deductible.

This program is subject to change or discontinuance without notice.

FIGURE 17.2 (continued)

HP Toner Cartridge Recycling Program

This program is in effect in the continental United States. For information regarding similar programs available in Hawaii, Alaska, Canada and Europe, please contact:

for Hawaii & Alaska   1-800-752-0900 Ext. 1872

for Canada-wide   1-800-387-3867 Dept. 129
  in Toronto, call: 678-9430 Ext. 4981

for other countries, please contact your HP toner cartridge supplier or your local HP sales office.

Hewlett-Packard Company
Customer Information Center
1-800-752-0900 Ext. 1872

National Wildlife Federation
1400 Sixteenth Street, N.W.
Washington, D.C.
20036-2266

The Nature Conservancy
1815 North-Lynn Street
Arlington, Virginia 22209

Printed on recycled paper.
PUB.F-IM-308 C   0891 F 650   Printed in Japan

FIGURE 17.2 (continued)

18

Staff Training

The standard puts considerable emphasis on the training of staff, in addition to establishing the level of competence of individuals to enable them to carry out their environmental tasks. It also envisages environmental duties being written into job descriptions.

18.1 Motivation and morale

ISO 9000 has demonstrated how staff can be motivated and morale boosted by the achievement of a top international quality standard, but the potential for staff satisfaction is even greater with BS 7750, as everyone likes to be seen to be working with an environmentally caring company. ISO 9000 has also demonstrated that the careers of staff are enhanced when they work for the kind of world-class employers who earn the top quality and environmental distinctions.

There should, therefore, be little problem in announcing that one's company is embarking upon a project to implement both BS 7750 and obtain a logo verifying the EC Eco-Management scheme. This announcement will help set in motion the project itself and smooth the way for the Preparatory Review and the implementation of procedures and controls.

What now follows is an outline of the different phases of training, together with some guidance on material.

18.2   The announcement

This is a sample letter to staff.

> ENVO Industries has made a policy decision to implement an environmental management system, based on BS 7750 and the EC Eco-Management scheme (or an EPA equivalent in non-European countries). When this is implemented by the end of the year, we will apply for registration to both BS 7750 and the EC Eco-Management scheme logo. This will be a great achievement and will demonstrate to our customers, our community and the world at large that ENVO Industries is an environmentally-caring company, accredited by an independent third party agency.
>
> This will also be a great tribute to all the staff in our team, all of whom are needed to carry out this policy, and, when we have reached the standard, we will have achieved something we can all be proud of. We will also be the first site in our group to be approved, which will be a great boost within the corporation worldwide, as other sites will then follow our example.
>
> You will shortly be hearing about how and when the project will affect you, and a training course will be introduced to ensure that you know how you can best contribute.
>
> Joe McCool, our Quality Manager, is in charge of the project, while the health and safety elements will be adminstered by Cathy West, our Human Resources Manager.
>
> I know I can expect you all to make a big effort to ensure that we achieve this great distinction.
>
> Signed
>
>
> Chief Executive

18.3 Who should do the training?

This is a very difficult question to answer as the standard is so new, and there are still few consultants who have much experience with it. If there is
a training capacity within the organization, then material such as this book may be used to develop internal courses. The following paragraphs contain suggestions about course content.

18.4 Training course content

While this book may be suitable for a quality manager or senior engineer facing the task of implementing the project, there is simply too much material here for a staff training course. One also needs a somewhat different emphasis for different levels of management, with a strong marketing and PR element at the top. Some of the chapter material, however, will be used in the proposed outline which follows. Firstly, however, a look at possible videos and other material.

18.5 Using videos and other materials

18.5.1 Outline of an in-house video

An in-house video has the advantages that it can be very specific to the operations of the company and offers a range of possibilities from the low budget and virtually one-off product to a good quality medium-term production. Any lack of quality will also be somewhat offset by the amusement of staff at seeing themselves and by the specific nature of the applications and processes which make it relevant.

If time and budget are available, a company can hire an expert outside production team for a quality product, but a simpler model can be made by insiders. The latter has one other advantage, especially if made over a period of months – an inventory of real situations can be put together. Here is a proposed outline.

1. Shoot a frank, unrehearsed and possibly unannounced appraisal of the company, from the state of the CEO's car and the front gardens to the quality of reception at the lobby and the refuse skip in the backyard.

2. Generate an inventory of possible non-conformities. Some companies may have mixed views about this, not wanting to consider a non-conformity, and the nature of some products may make it impossible. Good examples might be wrong settings on equipment, non-calibrated equipment, unsupervised situations.

3. The key documentation can easily be filmed. The material in this book can be used as a model. Typical documents could be:

   The Register of Regulations.
   Its title page.
   Its contents sheet.
   One page showing an issue, for example, Emissions.
   An actual regulation.
   Title pages of both the Register of Environmental Effects and the Environment Management Manual.
   Sample pages from inside.
   Control documents (for example, checklists, logs, monitoring sheets, graphs).

4. Breakdown of product. Take the product to pieces, physically if possible. Show the individual components, the raw materials it is made from, the packaging, and so on.

   Build a chart or show the hierarchy of actual components detailing place of origin (for example, timber from Canada, where there is real re-afforestation) and possible interactions with the environment. The video could even incorporate a discussion on each of the elements at this stage.

   Discuss sources of materials, effects on environment, effects of product use and final disposal.

5. Show all aspects of how the company interacts with the environment, which can be very visual.

6. Produce a separate section on health and safety, actually filming such activities as: walking, lifting, using equipment, wearing safety equipment and safety signs. Also, show aisles, lighting, access routes, and so on, in accordance with the checklists in this book.

18.5.2 Buying or renting a video

With new health and safety regulations in force and environmental demands increasing, there are now a number of good training videos on the market and in the pipeline. For a list of health and safety videos contact Gower, Gower House, Croft Road, Aldershot, Hampshire, GU11 3HR, UK.

18.5.3 Slides and overheads

A training session can also be greatly enhanced with slides and overheads. Each relevant issue, such as waste, packaging, and emissions should be photographed, as should each activity under the scrutiny of the health and safety regulations, and each element, such as lifting, VDU use, equipment, and workplace.

The major documents shown in this book make excellent overheads, which can be reproduced on a copying machine or on a laser printer. The documents recommended are those also recommended for the video: the Register of Regulations, its contents and sample pages, the title pages of the Register of Environmental Effects and the Environment Management Manual, contents and sample pages, and the checklists, logs, control documents, charts and graphs.

18.6 Using training to establish issues

18.6.1 Gathering facts

One characteristic of environmental management systems is that they are a continuing process, and therefore demand continuing appraisal, review and training. An excellent way to achieve relevance is to use the training process to gather information. This can be done by incorporating a module in which the participants contribute to the information-gathering process.

Here is a proposed approach. Ask each participant to describe his or her activity in terms of process, environmental results and potential with regard to damage to the environment and health and safety aspects. For example:

Activity	Environment result	Potential
Acid immersion	Used acid for disposal	Inhalation
		Scalding
		Blinding
		Leak to drains
		Fumes to air

Then take each potential event and analyse how it might happen.

Activity	Non-conformance	Was training given?	How monitored
Dipping	Not wearing safety mask	Yes	Not monitored

From this kind of analysis, a very useful picture can emerge of how good the current controls are and what changes need to be made. All such records from the training sessions should be passed to the quality manager for analysis and incorporation into the review mechanism. This whole process should be treated as if it were the result of an audit and should go through the normal review procedures. These sessions could also produce case studies for future use.

18.6.2 Case studies

Using the results obtained above, write up or photograph or film case studies. These can be projected under three broad headings:

1. The standard environment issues: air, water, noise and so on.
2. Accidents and emergency potential and procedures.
3. Health and safety issues.

This could be a very useful exercise which may reveal hitherto unknown information, but which may also satisfy both the training requirements of the standard and those of the health and safety regulations.

In the normal situation where serious incidents have not occurred, one may still cover potential accidents and emergencies.

18.7 Proposed outline for staff training course

Outline for one day orientation course in BS 7750.

9.30 am	Introduction

The new environment for business
- Regulations, i.e. environment, health and safety, product liability.
- Market demands.
- Quality standards.
- EC, CEN, EPA, ISO.
- A glance at our Register of Regulations.

10.30 am The issues for our organization
- Our product and its use.
- Our packaging.
- Emissions to the air.
- Water discharges.
- Materials and energy use.
- Solid and toxic waste.
- Noise and nuisance.
- Landscape, wildlife, amenity.
- Health and safety of employees and visitors.

11.30 am What is BS 7750?
- It is the new management standard.
- Relationship with ISO 9000.
- The elements involved.
- Costs and benefits.
- What the project will involve.

2.00 pm How we control the issues concerning us
- Defining the issues.
- Managing the issues and controls.

3.00 pm The environmental management system
- The checklists.
- The Environment Manual.

4.00 pm Practical examples

This framework could be adapted to form a two-day programme if more time is needed.

18.8   The training programme

The standard asks for a training programme along the following lines: training for executives, all operating personnel and new recruits; introductory and refresher programmes; recognition of performance; encouragement of employee suggestions and participation in environmental initiatives.

Issue Identification Summary

Envirocompany - Issue identification summary   For000

Initials:_____   Date:_____   Page 1 of 2

Issue	Yes	No
Planning:		
• New buildings		
• New Jetties		
• New Terminals		
• New Product outlets		
Legal requirements:		
• Conformance		
Product dispensing		
• Ergonomics		
Noise:		
Nuisance:		
Trees, amenities, landscape, and wildlife:		
• Trees		
• Amenities		
• Landscape		
• Wildlife		
Use of materials:		
• Raw materials source		
• Construction or manufacturing of raw materials		
Energy:		
• Energy source		
• Consumption		
Product life cycle:		
• Reusable		
• Disposable		
Product quality:		
• Environmental impact		
• Product constituents		
Product use:		
• Product handling		
• Product use		

Envirocompany - Issue identification summary   For000

Initials:_____   Date:_____   Page 2 of 2

Issue	Yes	No
Waste: • Storage • Removal/transport **Toxic waste:** • Storage • Removal/transport **Emissions:** • Chimney • Incinerator • Fumes **Effluent Discharge:** • Ground seepage into local aquifer/water table • Rivers • Marine • Sewage **Product transport, storage, and distribution:** • Transport • Storage • Distribution **Safety:** • Public • Staff (see Health & Safety check lists) • Customer (see product use) **Packaging:** • Package construction • Package and contents • Package life cycle		

Sample Register
of Regulations

Envirocompany
Register of Environment Regulations

This is a generic model suitable for a range of companies
from software development to processing.

Cover page   Page 1 OF 21

Rev: 001   Date:16/6/92

Document: Doc010

Quality/Environment Manager:

Description

This document contains the register of
regulations for the environment management
system and is a mandatory requirement of
the system.

Alterations are not permitted without prior
approval of the Quality Manager and must be
applied using the system for amendment
control contained within this document

Verification			
Description	Signature	Function	Date
Compiled by			
Approved by			
Authorised by			

Envirocompany Register of regulations	PAGE 2 OF 21
Amendment List	REV: 001 DATE: 16/6/92 Quality manager:

Amendment list

CHANGE REVISION NO.	DATE	BRIEF DESCRIPTION OF CHANGE	SECTION/S INVOLVED	PAGE NOs.
001	1/1/93	Update of Circulation list	1.7.0	1
002	3/4/93	Updating references to amendment list	1.1.0	1

Envirocompany Register of regulations	PAGE 3 OF 21
Environment regulations	REV: 001 DATE: 16/6/92 Quality manager:

Regulations:

EC Eco-scheme

Physical planning

Waste

Packaging

Visual display unit

Software ergonomics

Nuisance and noise

Trees, amenities, landscape, and wildlife

Emissions

Effluent discharges

Use of materials

Use of energy

Product life cycle

Envirocompany Register of regulations	PAGE 4 OF 21
Environment Relevant regulations	REV: 001 DATE: 16/6/92 Quality manager:

EC Eco management and auditing scheme

General

Regulation

The main overall environmental regulation is the EC Eco-scheme, in force since 1 January 1993. This is voluntary for participating companies for the first four years, but mandatory for member states in the setting up of support and accreditation procedures. It is a comprehensive system covering all aspects of our interaction with the environment.

Policy

It is our policy to comply right away with the Eco-scheme, to adopt the logo, and to be listed in the EC Official Journal as an Eco-scheme registered company.

We will achieve this by implementing BS 7750 as the main mechanism for meeting the requirements of the Eco-scheme.

Environment Effects Register reference:

Envirocompany Register of regulations	PAGE 5 OF 21
Environment Relevant regulations	REV: 001 DATE: 16/6/92 Quality manager:

Physical planning

Regulations

All physical planning of extensions to our facility or any new development work come under two groups of planning regulations. These are:

The local authority (Planning and Development) Act 1993 (No.28 of 1993) and all amendments since.

The Building Regulations   1991:

Policy

All physical planning and extension work at our facility will be carried out strictly in accordance with these regulations. No work will be authorized or commenced without the Quality Manager's approval, which will be given only where he/she is satisfied that the necessary controls to ensure that the work is done under the above regulations are in place. This, in turn, will be verified in writing by the local authority which will state that the work meets the approval of that authority under the above regulations.

Environment Effects Register reference:

For space reasons the Local Authority Act and Building Regulations are not included in this Register. It is however company policy that our architects ensure that they are complied with to the letter in any future planning or building work.

Envirocompany Register of regulations	PAGE 6 OF 21
Environment Relevant regulations	REV: 001 DATE: 16/6/92 Quality manager:

Waste

Regulations

There are two regulations applicable to our operations. These are:

European Communities (Waste) Regulations, 1979 (S.I. No 390 of 1979).

European Communities (Toxic and Dangerous Waste) Regulations, 1982 (S.I. No 33 of 1982)

Policy

There are also principles and policies laid out in the EC document A Community Strategy for Waste Management SEC (89) 934 1989. It is our policy to support the principles stated here in our operations, particularly in our influence over the design of our products and their packaging to reduce waste.

The first of the above regulations demands that local authorities be responsible for the provision of waste management in our area. It is our policy to commit all of our waste to the local authority, or to a registered operator, registered under the 1978 regulations. It is our policy to check that all waste disposal operators used by us produce evidence of such registration. For as long as the local authority, or registered operator, continue to manage our waste this policy will apply. In the event of the local authority, or registered operator, not being able to manage our waste, we shall store it safely on site until the local authority resumes management, or until safe and legal alternative management can be arranged.

In the case of toxic waste, we will also co-operate fully with the local authority, which also has responsibility for its management. In addition, we will endeavour in every way possible from design of product through to packaging to reduce or eliminate any toxic elements in our products.

Environment Effects Register reference:

APPENDIX 2

Envirocompany Register of regulations	PAGE 7 OF 21
Environment Relevant regulations	REV: 001 DATE: 16/6/92 Quality manager:

Packaging

Regulations

The relevant regulations and codes are:

The Green Dot Code (German)
The 4th and Final EC Directive on Packaging Waste
The EC Eco-labelling Directive.

Policy

It is our policy to conform to all of the above regulations and codes including the Green Dot Code in Germany, until it is replaced by EC legislation.

Environment Effects Register reference:

Envirocompany Register of regulations	PAGE 8 OF 21
Environment Relevant regulations	REV: 001 DATE: 16/6/92 Quality manager:

VDU Usage

Regulations

The standard behind the VDU Health and Safety regulation (which is for users), which applies to hardware and software manufacturers is:

ISO 9241 - Standard on the ergonomics requirements for work with Video Display Terminals. This applies to software as well as hardware.

ISO 9000 - Quality management standard. This together with BS 7750 helps to ensure our conformance to the above as well as the rest of our management system.

ISO 9001 - This standard helps to ensure that our software meets ISO 9241.

Policy

It is our policy to design and produce our software so that it conforms to the appropriate visual display requirements of ISO 9241. This is to ensure that users of our software will satisfy the software demands of the EC directive on VDUs.

It is also our policy to operate to ISO 9000 in the management of our software production.

Environment Effects Register reference:

Envirocompany Register of regulations	PAGE 9 OF 21
Environment Relevant regulations	REV: 001 DATE: 16/6/92 Quality manager:

Software Ergonomics

Regulations

ISO 9241 is the international standard for Ergonomics requirements for office work with visual display terminals (VDTs).

Its implementation by hardware and software manufacturers satisfies the demands of the EC directive on VDUs, but it goes further than VDUs and addresses the ergonomics of software itself. In this respect it creates an additional environmental demand on software producers - to ensure that their end product meets an international standard adopted or desired by users.

The section dealing with software ergonomics and man-machine interface specifies: dialogue principles, usability, presentation, user guidance and media dialogues.

ISO 9001 - This standard will help us to ensure delivery of product to ISO 9241.

Policy

It is our policy to design and develop our software products along the lines laid down by ISO 9241 including those dealing with software ergonomics and man-machine interface.

We will also manage our software production by a system meeting the requirements of ISO 9001 to ensure that product is delivered to ISO 9241.

Environment Effects Register reference:

Envirocompany Register of regulations	PAGE 10 OF 21
Environment Relevant regulations	REV: 001 DATE: 16/6/92 Quality manager:

Nuisances and Noise

Regulations

Only one regulation under nuisances and noise has relevance for us as we presently operate. This is the European Communities (Lawnmowers) (Permissable Noise Levels) Regulations, 1989 (S.I. No 102 of 1989)

Policy

The lawnmowers in use are supplied by .... and these mowers are certified to conform to the EC directive on lawnmowers. See document.

Inspections are carried out to ensure that the lawnmowers in use are the name and brand described in the document.

Environment Effects Register reference:

Envirocompany Register of regulations	PAGE 11 OF 21
Environment Relevant regulations	REV: 001 DATE: 16/6/92 Quality manager:

Trees, amenities landscape and wildlife

Regulations

There are a number of regulations in the area of trees, amenities, landscape and wildlife. It is doubtful if these will ever be infringed by our operations.

Policy

It is our policy to maintain and enhance the landscaping of our site at XXX. The grounds already exist to a high state of amenity. The site will be maintained and added to where required. No tree shall be cut down unless its condition, such as state of deterioration or danger to passers by or staff, warrants our doing so.

No wildlife whatsoever protected under the wildlife acts will be interfered with at our XXX site.

Environment Effects Register reference:

If found to be relevant at any stage the appropriate regulations will be filed in this Register.

Envirocompany Register of regulations	PAGE 12 OF 21
Environment Relevant regulations	REV: 001 DATE: 16/6/92 Quality manager:

Emissions

Regulations

Control of Atmospheric Pollution Regulations, 1990 (S.I. No 156 of 1990)

Air Pollution Act, 1992 (Air Quality Standards) Regulations, 1992 (S.I. No 244 of 1987)
Air Pollution Act, 1991 (Licensing of Industrial Plant) Regulations, 1991
This last act is the relevant one, giving general controls and licenses.

Local Authority (Planning and Development) General Policy Directive, 1988 (S.I. No. 317 of 1988) (Smokeless heating).

Policy

The Control of Atmospheric Pollution Regulations relates to industrial smoke of varying degrees of darkness and the time periods allowed for emissions.

The Air Pollution Act, 1992, relates to emissions of sulphur dioxide, suspended particulates, lead and nitrogen dioxide.

The Air Pollution Act 1991, (Licensing of Industrial Plant) Regulations 1991 specifies actual permissable emissions.

The Local Government General Policy directive relates to smokeless fuel for heating.

Our heating fuel is at all times smokeless in the sense of the last directive, which bans non-smokeless fuel use in certain built up areas.

The first two regulations apply to our boiler house. It is our policy to ensure that our boiler house controls implement these regulations as a baseline, beyond which further performance targets will be set.

Environment Effects Register reference:

Envirocompany Register of regulations	PAGE 13 OF 21
Environment Relevant regulations	REV: 001 DATE: 16/6/92 Quality manager:

Emissions

<u>Regulations</u>

The same as above, but this is now for a processing company, which has licensed emissions.

We are licensed by the Local Authority to emit the following emissions in the amounts stated.

Emissions	Amounts licensed
Particulates	At this stage we simply state the local authority
SOx	requirement
NOx	
HCl	
CO	

Environment Effects Register reference:

Envirocompany Register of regulations	PAGE 14 OF 21
Environment Relevant regulations	REV: 001 DATE: 16/6/92 Quality manager:

Effluent discharges

Regulations

The relevant regulations are

Local Authority (Water Pollution) Regulations 1988 (S.I. No. 108 of 1988.)
Local Government (Water Pollution) Amendment Act 1990

Local Government (Water Pollution) (Control of Cadmium Discharges) Regulations, 1985 (S.I. No. 294 of 1985)

Local Government (Water Pollution) Act 1989 (Control of Hexachlorocyclohexane and Mercury Discharges) Regulations, 1989 (S.I. No. 55 of 1989).

The Cadmium regulations give effect to EC Directive 83/513/EEC of 26 September 1983.

The Mercury and Hexachlorocyclohexane regulations give effect to EC Directives 84/419/EEC of 9 October 1984 and 84/156/EEC of 8 March 1984.

Policy

It is company policy to ensure that all effluent discharges conform to the above regulations.

As far as the cadmium, hexachlorocyclohexane and mercury discharges regulations are concerned, this company will not be discharging these substances in any quantities whatsoever into the waste water system.

Environment Effects Register reference:

Envirocompany Register of regulations	PAGE 15 OF 21
Environment Relevant regulations	REV: 001 DATE: 16/6/92 Quality manager:

Effluent discharges

Regulations

The same as the above, but this is now for a company which
has effluent discharges licensed by the local authority.

Discharge	Amount licensed
BOD	At this stage we simply
SS	state the local authority
COD	requirement
Oils	
MBAs	
SO_4	
Cl	
NH_3	
NO_3	
PO_4	
Metals	

Environment Effects Register reference:

Envirocompany Register of regulations	PAGE 16 OF 21
Environment Relevant regulations	REV: 001 DATE: 16/6/92 Quality manager:

Use of Materials

Regulations

While no specific regulations cover use of materials, different materials may attract individual regulations. The use of materials is also a fundamental aspect of environmental control. As regulations emerge affecting the materials we use, they will be incorporated here and complied with; meanwhile the policy statement below records our objectives in this area.

Policy

It is our policy to use only materials not restricted or prohibited by regulations and to use all materials as sparingly and efficiently as possible. Our ISO 9000 quality management system will be maintained as the most efficient means of reducing waste.

Our policy on energy usage is shown separately, but water will also be treated as a resource to be cared for and it is our policy to ensure that water is not wasted. All local authority restrictions on water during periods of shortage will be strictly complied with.

Environment Effects Register reference:

Envirocompany Register of regulations	PAGE 17 OF 21
Environment Relevant regulations	REV: 001 DATE: 16/6/92 Quality manager:

Use of energy

Regulations

No specific regulations apply to our use of energy, but energy use is
an important environmental matter. Regulations however may apply from
time to time during periods of energy shortage, and there are
continuous national and international campaigns of energy reduction in
industry.

Policy

It is our policy to maintain an energy conservation programme within
the plant on a long-term basis. This includes being aware of all
energy conservation possibilities and operating a set of guidelines
for the control of energy use within the plant. These guidelines are
laid down in the Register of Environmental Effects.

Environment Effects Register reference:

Envirocompany Register of regulations	PAGE 18 OF 21
Environment Relevant regulations	REV: 001 DATE: 16/6/92 Quality manager:

Product life cycle

<u>Regulations</u>

While no specific regulations apply to our product life cycle, different pieces of legislation can from time to time be related to the materials we use and to their disposal. These can include both main products, their use in the field, and packaging.

<u>Policy</u>

It is our policy to manufacture and procure as far as possible all of our products and their packaging in an environmentally sound manner. This policy ranges from design and production through packaging and localization to final end use. To this end we have analysed our product in the Register of Environmental Effects to establish the environmental issues involved and to ascertain where and how we can take steps to ensure that our product is as environmentally sound as possible.

Environment Effects Register reference:

Envirocompany Register of regulations	PAGE 19 OF 21
Environment Actual regulations	REV: 001 DATE: 16/6/92 Quality manager:

Here we file the actual regulations and directives by punching holes
in them or, better still, inserting them into pre-punched plastic see-
through envelopes into the ring binder.
This will mean producing a fairly hefty register.

Sample directive

Specimen

STATUTORY INSTRUMENTS.

S.I. No. 266 of 1993

AIR POLLUTION ACT, 1992, (LICENSING OF
INDUSTRIAL PLANT) REGULATIONS, 1993

(PI.5942)

2 [266]

S.I. No. 266 of 1993.

AIR POLLUTION ACT, 1992,(LICENSING OF
INDUSTRIAL PLANT) REGULATIONS,1993.

The Minister for the Environment, in exercise of the
powers conferred on him by sections 10,30,31,33,34 and

35 of the Air Pollution Act, 1992 (No. 6 of 1992)
hereby makes the following Regulations:-

Citation.   1. These Regulations may be cited as the Air
Pollution Act, 1992 (Licensing of Industrial Plant),
Regulations,1993.

Commencement. 2. These Regulations shall come into operation on the
1st day of November, 1993

Envirocompany Register of regulations	PAGE 20 OF 21
Health and safety regulations	REV: 001 DATE: 16/6/92 Quality manager:

For space reasons all the health and safety regulations and policy statements will not now be listed. A summary follows with one example. Simply repeat for these issues the format above, one sheet for each showing regulation and policy and at the end of these sheets all the actual regulations in see-through envelopes inserted into the ring binder. The issues are:

Framework

Workplace

Work equipment

Visual display units

Manual handling

Personal protective equipment

Pregnant workers

Signs

Envirocompany Register of regulations	PAGE 21 OF 21
Health and safety Relevant regulations	REV: 001 DATE: 16/6/92 Quality manager:

Health and Safety   One sample. Draw up one sheet for each issue.

<u>Regulations</u>

Framework
The EC Framework Directive 1992 on health and safety

<u>Policy</u>

Thirteen work place elements are set out for attention in the
workplace regulations. All of these are set out in the Environmental
Effects Register and the procedures to ensure that they are met are
laid down in the Environment Management System. All are filed also in
this register in the section which follows.

It is our policy to ensure that all the thirteen demands of the
regulations are met by us.

In addition to the workplace and the other specific Health and Safety
regulations which follow, the Framework directive sets the overall
policy and framework for health and safety. It is company policy to
conform to this by conforming to the specific regulations here and
following.

A separate Safety Statement as required by the 1989 Act has been drawn
up. See Document number (or included at the back of this Register).

Environment Effects Register reference:

Sample Register of Environmental Effects

Envirocompany
Register of Environmental Effects

Cover page	Page 1 OF 15
Rev: 001	Date:16/6/92
Document: Doc010	
Quality/Environment Manager:	

Description

This document contains the register of effects for the environment management system and is a mandatory requirement of the system.

Alterations are not permitted without prior approval of the Quality Manager and must be applied using the system for amendment control contained within this document

Verification			
Description	Signature	Function	Date
Compiled by			
Approved by			
Authorised by			

Envirocompany Register of effects	PAGE 2 OF 15
Amendment List	REV: 001 DATE: 16/6/92 Quality manager:

Amendment list

CHANGE REVISION NO.	DATE	BRIEF DESCRIPTION OF CHANGE	SECTION/S INVOLVED	PAGE NOs.
001	1/1/92	Update of Circulation list	1.7.0	1
002	3/4/92	Updating references to amendment list	1.1.0	1

Envirocompany Register of effects	PAGE 3 OF 15
Environment effects	REV: 001 DATE: 16/6/92 Quality manager:

Issues:

EC Eco management and auditing scheme

Physical planning

Waste

Packaging

Nuisance and noise

Trees, amenities, landscape, and wildlife

Emissions

Effluent discharges

Use of materials

Use of energy

Product quality

Envirocompany Register of effects	PAGE 4 OF 15
Environment Relevant effects	REV: 001 DATE: 16/6/92 Quality manager:

EC Eco management and auditing scheme

General

Issue

The main overall environmental regulation is the EC Eco-scheme, which came into force on 1 January 1993. This is voluntary for participating companies for the first four years, but mandatory for member states in the setting up of support and accreditation procedures. It is a comprehensive system covering all aspects of our interaction with the environment.

The Envirocompany situation

We are in a position to implement our environmental management system in a manner which should satisfy both the EC Eco-scheme and BS 7750.

Policy and targets
It is our policy to implement and maintain an environmental management system along the lines of BS 7750 to satisfy the requirements of the EC Eco-scheme. Until such time as a local Eco certification scheme is in place, and after we achieve BS 7750, it will be our policy to maintain our BS 7750 status, and to promote ourselves accordingly. If and when we achieve the EC Eco-scheme we will add that logo also and promote our status as having achieved both.

Register of Regulations reference:

Envirocompany Register of effects	PAGE 5 OF 15
Environment Relevant effects	REV: 001 DATE: 16/6/92 Quality manager:

Physical planning

<u>Issues</u>

All physical planning of extensions to the Envirocompany
facility at XXX or any new development work come under two
groups of planning regulations. These are:

The local Government (Planning and Development) Act 1983
(No.28 of 1963) and all amendments since.

The Building Regulations 1991:

<u>The Envirocompany situation</u>

At the time of installing the system there are no
extensions planned to our facility, so there are no
planning issues.

<u>Policy and targets</u>

It is our policy that all extensions, changes or new
building projects at our XXX site will be done strictly in
accordance with the Local Government Act and Building
Regulations as described in the Physical Planning section
of the Register of Regulations.

Register of Regulations Reference:

Envirocompany Register of effects	PAGE 6 OF 15
Environment Relevant effects	REV: 001 DATE: 16/6/92 Quality manager:

Waste

Issues

There are two regulations applicable to our operations.
These are:

European Communities (Waste) Regulations, 1979 (S.I. No 390
of 1979).

European Communities (Toxic and Dangerous Waste)
Regulations, 1982 (S.I. No 33 of 1982)

The Envirocompany situation

We use XYZ (Supply name of waste disposal operator here) to
dispose of all of our waste. XYZ is registered by Durham
County Council to transport and dispose of waste. We
control this issue through Procedure Number - YYY.

In the matter of toxic waste our products contain (specify
here if relevent). We are minimizing any adverse effects of
the careless disposal of used products by our customers by
inserting instructions in our packs as shown in Procedure
WWW.

Policy and targets
It is our policy to use only waste disposal contractors
authorised by the local authority to handle our waste. We
control this issue through the controls in the above
procedure.

In the matter of the toxic content of our products we
control this issue through the Procedure shown at JJJ.

Register of Regulations Reference:
Environment Manual Reference:

197

Envirocompany Register of effects	PAGE 7 OF 15
Environment Relevant effects	REV: 001 DATE: 16/6/92 Quality manager:

Packaging

Issues

The relevant regulations and codes are:

The Green Dot Code (German)
The 4th and Final EC Directive on Packaging Waste (about to
be published)
The EC Eco-labelling Directive (underway).

The Envirocompany situation

We deal with the German market as follows (specify here).

As there are still no specific regulations relating to our
packaging, the issues for us are that we conform to the
general aspirations and emerging codes of practice
concerning packaging. The main issues of relevance to us
are that we not use environmentally-undesirable materials
in our packaging and that we do not overpack products.
These are procurement issues which we have taken up with
our suppliers and control in Document PPP.

(We need to ask suppliers to tell us the latest codes of
good practise so that we can design controls)

Policy and targets
Our policy is to conform as far as possible with current
best environmental practises in packaging and to also
conform fully with all legislation concerning packaging

Register of Regulations Reference:
Environment Manual Reference:

Envirocompany Register of effects	PAGE 8 OF 15
Environment Relevant effects	REV: 001 DATE: 16/6/92 Quality manager:

Nuisances and noise

Issues

Only one regulation under nuisances and noise has relevance
for us as we presently operate. This is the European
Communities (Lawnmowers) (Permissable Noise Levels)
regulations, 1989 (S.I. No 102 of 1989)

The Envirocompany situation
Our lawns are maintained by XXX, whom we have advised to
conform to the above regulation. This is controlled by
Document MMM.

Policy and targets
It is our policy to ensure that all our suppliers providing
lawn mowing services use machinery conforming to the above
regulation.

Register of Regulations Reference:
Environment Manual Reference:

Envirocompany Register of effects	PAGE 9 OF 15
Environment Relevant effects	REV: 001 DATE: 16/6/92 Quality manager:

Trees, amenities landscape and wildlife

Issues

There are a number of regulations in the area of trees, amenities, landscape and wildlife. It is doubtful if these will ever be infringed by our operations.

The Envirocompany situation

Our XXX site is landscaped to a high degree and there should be no issues here to be dealt with. Only in the event of storm damage or disease or danger of falling will removal of material other than normal trimming or cutting back take place. This is not an issue needing formal controls.

Policy and targets

It is our policy to maintain our site to a constant high standard of landscaping.

Register of Regulations Reference:
Environment Manual Reference:

Envirocompany Register of effects	PAGE 10 OF 15
Environment Relevant effects	REV: 001 DATE: 16/6/92 Quality manager:

Emissions

<u>Issues</u>

Control of Atmospheric Pollution Regulations, 1970 (S.I.
No 156 of 1987)

Air Pollution Act, 1987 (Air Quality Standards)
Regulations, 1987 (S.I. No 244 of 1987)
Air Pollution Act, 1987 (Licensing of Industrial Plant)
Regulations, 1988
This last act is the relevant one, giving general controls
and licenses.

Local Government (Planning and Development) General Policy
Directive, 1988 (S.I. No 317 of 1988) (Smokeless heating).

<u>The Envirocompany situation</u>

Emissions	Amounts licensed	Actual outputs
Particulates		
SO_x		
NO_x		
HCl		
CO		

<u>Policy and targets</u>
It is our policy to ensure that we continue to maintain our
outputs within the targets as shown above.

Register of Regulations Reference:
Environment Manual Reference:

Envirocompany Register of effects	PAGE 11 OF 15
Environment Relevant effects	REV: 001 DATE: 16/6/92 Quality manager:

Effluent discharges

<u>Issues</u>

The relevant regulations are

Local Government (Water Pollution) Regulations 1978 (S.I.
No. 108 of 1978.)
Local Government (Water Pollution) Amendment Act 1990

Local Government (Water Pollution) (Control of Cadmium
Discharges) Regulations, 1985 (S.I. No. 294 of 1985)

Local Government (Water Pollution) Act 1977 (Control of
Hexachlorocyclohexane and Mercury Discharges) Regulations,
1986 (S.I. No. 55 of 1986).

The Cadmium regulations give effect to EC Directive
83/513/EEC of 26 September 1983.

The Mercury and Hexachlorocyclohexane regulations give
effect to EC Directives 84/419/EEC of 9 October 1984 and
84/156/EEC of 8 March 1984.

<u>The Envirocompany situation</u>

Discharge	Amount licensed	Actual outputs
BOD		
SS		
COD		
Oils		
MBAs		
SO_4		
Cl		
NH_3		
NO_3		
PO_4		
Metals		

Policy and targets

It is our policy to ensure that we continue to maintain our
outputs within the targets as shown above.

Register of Regulations Reference:
Environment Manual Reference:

Envirocompany Register of effects	PAGE 12 OF 15
Environment Relevant effects	REV: 001 DATE: 16/6/92 Quality manager:

Use of materials

Issues

While no specific regulations cover use of materials,
different materials may attract individual regulations.
The use of materials is also a fundamental aspect of
environmental control.

The Envirocompany situation

Despite our having no specific materials regulations, to
meet current codes of good practice we must ensure that we
do not waste materials. The Preparatory Review revealed the
following areas where a programme of materials reduction
could reduce materials use.

(It is necessary to audit materials in use to look for
wastage)

Our Paper Recycling control is shown in Document RRR.

Policy and targets

It is our policy to (fill in after review).

Register of Regulations Reference:
Environment Manual Reference:

Envirocompany Register of effects	PAGE 13 OF 15
Environment Relevant effects	REV: 001 DATE: 16/6/92 Quality manager:

Use of energy

<u>Issues</u>

No specific regulations apply to our use of energy, but
energy use is an important environmental matter.
Regulations, however, may apply from time to time during
periods of energy shortage, and there are continuous
national and international campaigns of energy reduction in
industry.

<u>The Envirocompany situation</u>
All our energy use is controlled in our Energy Conservation
Programme contained in Document EEE.

<u>Policy and targets</u>

It is our policy to maintain and improve the Energy
conservation programme shown in the Environment Management
Manual.

Register of Regulations Reference:
Environment Manual Reference:

Envirocompany Register of effects	PAGE 14 OF 15
Environment Relevant effects	REV: 001 DATE: 16/6/92 Quality manager:

Product quality   (for a hardware manufacturer)

Issues

There are three regulations and standards relating to
product quality.
The first is compulsory, while the second and third help
support the first and other codes of practice.

The three are:

- The EIC standard on electricity safety
- ISO 9241 - the VDU ergonomics standard
- ISO 9001 - the ISO 9000 standard for managing design
  and manufacture
(There will be many more for a hardware manufacturer,
 including for example telecom regulations)

The Envirocompany situation

In the case of a subsidiary or a distributor the following
statement could be appropriate:

"These are all supplier issues concerning our parent
company and sole supplier. These are so important that
they are contained and controlled in a separate Product
Quality System - Document PPP."

Policy and targets

(For a subsidiary as above)

It is our policy to request our parent company in Maryland
to conform to the above regulations and standards and to
implement a system of monitoring and controlling their
delivery to us in relation to these issues.

Register of Regulations Reference:
Environment Manual Reference:

Envirocompany Register of effects	PAGE 15 OF 15
Health and Safety	REV: 001 DATE: 16/6/92 Quality manager:

As with the Register of Regulations, repeat the health and
safety issues here in the same manner as the foregoing
environmental issues.
For space reasons they will not be listed.

There is a difference, however, between these and the
environment issues. The former depend on operations and can
be quite variable, while these are fixed - there are no
variations in the requirements.

It should be acceptable, therefore, to simply state here
that the health and safety issues are all being met fully,
and to refer to the procedures where they are being
controlled, by reference to the Environment Management
Manual.

Register of Regulations Reference:
Environment Manual Reference:

Sample Environment Management Manual

Envirocompany
Environment Manual

Cover page	Page 1 OF 1
Rev: 001	Date:16/6/92
Q.A./Environment Mgr.: J. Green	

Description

This manual describes the Envirocompany environmental management system and is a mandatory requirement of the system.

Alterations are not permitted without prior approval of the Quality Manager and must be applied using the system for amendment control contained within this document

Verification			
Description	Signature	Function	Date
Compiled by			
Approved by			
Authorised by			

Envirocompany Environment manual	SECTION: 1.0.0 PAGE 1 OF 1
Table of Contents	REV: 001 DATE: 16/6/92 Quality manager: J. Green

Section	No. of sheets
1.0.0 TABLE OF CONTENTS	1
1.0 AMENDMENT LIST	1
2.0 AMENDMENT PROCEDURES	1
3.0 CIRCULATION LIST	1
4.0 DESCRIPTION OF COMPANY	1
5.0 PLACE OF ENVIRONMENT MANUAL IN OVERALL PROCEDURES	
6.0 POLICY	2
7.0 OBJECTIVES	
8.0 TARGETS	
9.0 KEY ROLES AND RESPONSIBILITIES	1
2.0.0 ENVIRONMENT MANAGEMENT SYSTEM	
1.0 CROSS REFERENCES	
REGISTER OF REGULATIONS 4.32	
REGISTER OF EFFECTS 431	
PROCEDURES	
ENVIRONMENT MANUAL 4.34	
EC ECO-SCHEME	

2.0   CHECK LISTS (after development of the Register of Environmental
              Effects these form the basis for the Environment
                        Management System)	

3.0 NON-CONFORMANCE 4.52	
3.1 WITHIN LIMITS	
3.2 INCIDENTS	
3.3 EMERGENCIES	
3.0.0 AUDITS AND REVIEWS	
4.0.0 TRAINING	

Envirocompany Environment manual	SECTION: 1.1.0 PAGE 1 OF 1
Amendment List	REV: 002 DATE: 16/6/92 Quality manager: J. Green

1.1.0 Amendment list

CHANGE REVISION NO.	DATE	BRIEF DESCRIPTION OF CHANGE	SECTION/S INVOLVED	PAGE Nos.
001	1/1/92	Update of Circulation list	1.7.0	1
002	3/4/92	Updating references to amendment list	1.1.0	1

Envirocompany Environment manual	SECTION: 1.2.0 PAGE 1 OF 1
Amendment procedures	REV: 001 DATE: 16/6/92 Quality manager: J. Green

1.2.0   Amendment procedures

The latest revision numbers of the manual are on the wall of the Quality
Manager's office as well as in the amendment list in Section 1.1.0. The
only valid copy of this manual is that shown with the latest revision
numbers.

1.2.1 All copies of the manuals and all revisions and additions
are controlled by him/her.

1.2.2 Changes and additions can be suggested by all staff
members and co-ordinated through appropriate managers. All final changes
must be carried out with the authority of the Quality Manager.

1.2.3 All changes and amendments are recorded on the Amendments
List (1.1.0). This list and all amended pages are then circulated to the
holders of each Environment manual. Holders must insert new pages and
destroy old. The Quality Manager may inspect manuals at any time.

Envirocompany Environment manual	SECTION: 1.3.0 PAGE 1 OF 1
Circulation list	REV: 001 DATE: 16/6/92 Quality manager: J. Green

1.3.0 Circulation list

This manual must be strictly controlled and maintained as a confidential document. It may be circulated only to those shown below:

Copy number	Holder	Title
1		Chief Executive
2	J. Green	Technical Manager
3	XXXXX	Financial Controller
4	J. Blow	
5	J. Hope	
6	S. Stoff	Staff Member
7	K Customer 1	Key Customer #1
8	K Customer 2	Key Customer #2
9	K Customer 3	Key Customer #3
10	A. Auditor	Auditor/Inspector

Envirocompany Environment manual	SECTION: 1.4.0 PAGE 1 OF 1
Description of company	REV: 001ᵛ DATE: 16/6/92 Quality manager: J. Green

1.4.0 Description of company

Envirocompany is based in XXX and supplies a wide range of - - -.

        To be filled in by company

Envirocompany Environment manual	SECTION: 1.5.0 PAGE 1 OF 1
Place of environment manual in overall procedures	REV: 001 DATE: 16/6/92 Quality manager: J. Green

1.5.0 Place of environment manual in overall procedures

This manual is part of the company's overall procedures.

1.5.1 This manual describes the procedures for operating and maintaining the company's environment management system.

1.5.2 The manual is strictly controlled by circulation and amendment. Only the latest issue number is valid. The Quality Manager controls all revisions, issues, and circulation. Only he or she decides what manuals will be in circulation.

1.5.3 All copies of this manual are numbered and all pages and copies are subject to control by the Quality Manager.

1.5.4 All of the procedures in this manual have been approved by management and express the environment policy laid down by management.

Envirocompany Environment manual	SECTION: 1.6.0 PAGE 1 OF 1
Policy	REV: 001 DATE: 16/6/92 Quality manager: J. Green

1.6.0 Environment policy statement

At a meeting on xxxx date management agreed to install an environment management system, and made the following Management Policy Commitment.

1.6.1 The Management of Envirocompany has adopted a policy of operating the service under control of an environment management system, installed and operated along the lines laid down in the environment management standard BS 7750. It is company policy to operate continuously to these standards, as they apply, and to seek annual registration from BSI assessors or local assessors when available.
It is also our policy to use our system to qualify for the EC Eco-scheme logo and to achieve a placement on the EC-wide published list of registered Eco-scheme companies.

1.6.2 The purpose of this policy is to ensure that the company operates to the stringent standards set by both BS 7750 and the EC Eco-scheme.
1.6.3 It is also our policy to continuously assess our environment procedures to look for improvements.

Envirocompany Environment manual	SECTION: 1.7.0 PAGE 1 OF 1
Objectives	REV: 001 DATE: 16/6/92 Quality manager: J. Green

1.7.0 Objectives

The objectives of the system are for Envirocompany to operate as
an environmentally-caring company and to demonstrate that achievement
through the employment of both BS 7750 and the EC Eco-scheme.

Envirocompany Environment manual	SECTION: 1.8.0 PAGE 1 OF 1
Targets	REV: 001 DATE: 16/6/92 Quality manager: J. Green

1.8.0 Targets

The targets are set out both in the Register of Regulations and the
Register of Environmental Issues. As we progress we expect to improve our
targets beyond those set by regulation.

Envirocompany Environment manual	SECTION: 1.9.0 PAGE 1 OF 1
Key roles and responsibilities	REV: 001 DATE: 16/6/92 Quality manager: J. Green

1.9.0 Key roles and responsibilities

The Quality Manager, reporting to the Chief Executive, has total authority for environment management and the full backing of the Chief Executive and management for all actions he may deem necessary in carrying out his job. In no circumstances may his decisions about the Environment Management System be over-ridden.

The Quality Manager is also responsible for the control of amendment to, and copies of, all environment management system documentation.

1.9.1 The environment control organization is as follows:

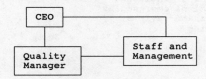

1.9.2 Responsibilities

Environment Assurance
The responsibilities of the Quality Manager include:

Establishing and documenting the environment standards and levels to be applied for each product and process. This will be clearly documented in the system and will include details of primary inspection points and control processes in the plant.

Ensuring that each department's responsibility for environment is documented in an acceptable format.

Ensuring that adequate records are maintained for demonstration of conformance to specified requirements.

Conducting internal audits on the system of environment assurance to ensure continued adherence to documented requirements.

Resolving non-conformances within the environment system.

Ensuring adequate resources are available to maintain the environment system at its required level.

Ensuring that the health and safety regulations are also catered for in the environment management system.

215

APPENDIX 4

Envirocompany Environment manual	SECTION: 2.0.0 PAGE 1 OF 1
Environment management system	REV: 001 DATE: 16/6/92 Quality manager: J. Green

2.0.0 Environment management system

The main elements covered in the system are as follows:

Policy
Targets
Organization
Register of regulations
Register of effects
Procedures
Environment management manual
Control of non-conformances
Control of incidents
Control of emergencies
Audits and reviews
Training

Envirocompany Environment manual	SECTION: 2.1.0 PAGE 1 OF 1
Cross references	REV: 001 DATE: 16/6/92 Quality manager: J. Green

2.1.0 Cross references

Document	Document No.
See document Register of Regulations	010
See document Register of Effects	011
See document Environment Procedures	012-015
See Operational Procedures	020-040
See document Environment Management Manual	001
See document EC Eco-scheme	009

Envirocompany Environment manual	SECTION: 2.2.0 PAGE 1 OF 15
Checklists	REV: 001 DATE: 16/6/92 Quality manager: J. Green

2.2.0 Checklists

The Checklists are used in the first instance in the Preparatory Review and later become the basis for the control procedures.

Checklists have been mentioned throughout this book, but one cannot give a definitive list as they will vary from company to company. To summarize once more, they are derived in the first instance by analyzing the actual regulations in the Register of Regulations as a means of auditing the potential issues needing control. As a result of the Preparatory Review and the construction of the Register of Issues, they can be converted into control documents.

This section could simply contain cross references to the control documents. The checklists are shown as they are a simple form of controls. Only some checklists are shown for space reasons. The health and safety check lists are quite easy to devise as the regulations are very specific.

Envirocompany - Checklist (Environment)

Initials:_____   Date:_____   For000
Page 1 of 1

Element	Action	Amt.	Lmt.
Emissions			
Particulate			
SO_x			
NO_x			
HCl			
CO			

Envirocompany - Checklist   (Environment)

Initials:_____   Date:_____

Element	Action	Amt.	Lmt.
Effluent discharge			
Particulate			
BOD			
SS			
COD			
OILS			
MBAs			
SO_4			
Cl			
NH_3			
NO_3			
PO_4			
Metals			

Envirocompany - Checklist   (Environment)

Initials:_____   Date:_____   For000
                                                       Page 1 of 1

Element	Action	Pass	Fail
Water resources (rivers)	Check: • Is river water used? • How much of overall flow?		
Water resources (rivers access aquatic wildlife)	Check: • Is access to upper and or lower parts of river effected?		
Water resources (rivers access public and other life)	Check: • Access to river effected?		
Water resources (rivers natural state	Check: • Down stream of process has condition and make up of river changed?		
Other resources (ie: predators)	Check: • Is control of predators having harmful effects?		

221

APPENDIX 4

Envirocompany - Checklist (Environment)

Initials:_____   Date:_____

For000
Page 1 of 1

Issue	Action	Pass	Fail
Package and contents	Check: • Define and identify • Redundancy • Non-biodegradable • Toxic • Reusable • Recyclable • Retrieve recycle • Alternate use after delivery • Monitor quantities		
Transport package	Check: • Reusable		

Envirocompany - Checklist (Environment)

Initials:_____   Date:_____   For000
Page 1 of 1

Element	Action	Pass	Fail
Toxic waste Special waste plan	Check: • does it exist?		
Register of waste production and movement	Check: • does it exist ?		
Transport of waste by a person other then the local authority	Check: • Valid permit • Clearly labelled with type of waste, waste control number, and source address name		
Storage of waste by a person who does not have a permit	Check: • Waste will be removed by a person holding a permit as soon as possible		
Control of waste storage	Check: • All waste will be marked appropriately • All waste will be stored in a designated area • Waste cannot escape into non-appropriate, skips, drains, or compacters etc.		

223

Envirocompany - Checklist (Environment)

Initials:_____   Date:_____

For000
Page 1 of 1

Element	Action	Pass	Fail
Waste Transport of waste by a person other tnen the local authority	Check: • Valid permit		
Storage of waste by a person who does not have a permit	Check: • Waste will be removed by a person holding a permit as soon as possible		
Control of waste storage	Check: • All waste will be marked appropriately • All waste will be stored in a designated area • Waste cannot escape into non-appropriate, skips, drains, or compacters etc.		

Shown are some samples for lifting and personal protective equipment.

Envirocompany - **Manual handling checklist** (Health & safety)

Initials:_____   Date:_____   For000
Page 1 of 5

Issue	Action (Checklist)	Pass	Fail
Overall physical workload unreasonably heavy? Does task require a weaker person to exert him/herself to the limit of his/her strength?	Check: • Lifting • Pushing • Pulling • Other		
Does task require a person to support a load or exert a sustained force for more than short period?	Check: • Lifting • Pushing • Pulling • Other		
Does lifting action commence in an unsatisfactory position?	Check: • Bend from waist/hips • Twist trunk • Lean over or reach out sideways		
Is lifting action performed outside the range of heights which is acceptable, given the nature of the load?	Check: • Necessary to lift above shoulder hight • Necessary to lift below knee hight • Necessary to make awkward movements to complete the lift • Necessary to handle heavy or difficult loads outside the preferred range of Heights as above		

225

Envirocompany - Manual handling checklist (Health & safety)

Initials:_____   Date:_____   For000
Page 2 of 5

Issue	Action (Checklist)	Pass	Fail
Lifting action involving hazardous movements	Check: • Turning motion • Sidestepping motion • Change of grip		
Does task require reaching actions which are beyond the cap- of a short person?			
Necessary to work long periods in an unsatisfactory posture	Check: • Stooped position • Crouching position • Hands above mid-torso Height • Twisted position • Asymmetric or side bending position • Demands of task prevent the worker from changing posit- at will • Worker required to take weight on one leg		
Repetition rates high	Check: • Necessary to perform more than 10 handling actions per minute • Necessary to handle heavy loads more than once per minute		
Does person work continuously without adequate rest periods?			
Does the person perform the same task continuously throughout the working day?			
Does weight of the load seem unreasonably heavy, considering the strength and fitness of the workforce?			

Envirocompany - Manual handling checklist (Health & safety)

Initials:_____   Date:_____   For000
Page 3 of 5

Issue	Action (Checklist)	Pass	Fail
Load difficult to handle	Check: • Too bulky • Too long • Too wide • Difficult to grip firmly • Unstable • Contents likely to shift during movement • Sharp edges, splinters etc • Trap fingers • Obstruct persons's vision • Hazardous or offensive in some other way		
Is there assistance available when handling difficult or heavy loads?			
Are there obstructions in the working area which will prevent the worker from keeping the load close to the body throughout the lift?	Check: • Necessary to reach over obstructions • Necessary to reach into containers • Worker free to adopt the most advantageous foot placements • Heavy or frequently used objects stored in inaccessible places		
Obstacles or hazards in working area	Check: • Things to bump into • Things to trip over • Steps or changes of level • Floor slippery • Floor needs cleaning • Rubbish or clutter that needs cleaning up		

227

Envirocomany - Manual handling checklist (Health & safety)

Initials:_____   Date:_____   For000
Page 4 of 5

Issue	Action (Checklist)	Pass	Fail
Problems of access or clearance for a tall or bulky person	Check: • Headroom • Passageways between obstacles • Legroom under working surfaces • Toe recesses under working surfaces		
Environmental problems	Check: • Lighting levels • Heat • Humidity • Cold • Wind • Ventilation • Dust • Noise		
Workers carrying out manual tasks	Check: • Less than 18 years • Greater than 55 years		
Do any workers have?	Check: • Limited range of motion in limbs and/or back • Heart or respiratory problems • History of hernias • Musculoskeletal disorder especially a previous history of back pain • Temporary impairment or debility		
Are any workers pregnant?			
Does any worker have specific difficulties in carrying out the manual handling tasks allotted to them?			

Envirocomapny - Manual handling Checklist (Health & safety)

Initials:_____   Date:_____   For000
                                                        Page 5 of 5

Issue	Action (Checklist)	Pass	Fail

Are there problems with regard to clothing or personal equipment?	Check:
	• Clothing hamper or constrain worker's movements
	• Items of safety clothing:
	  Required but not readily available
	  Readily available but not appropriate
	  Both readily available and appropriate but not used
	• Gloves
	• Boots
	• Headgear
	• Eye protection
	• Other		

Is there anything else which makes the task more difficult or hazardous than it needs to be?			

Envirocompany - P. P. E. checklist (Health & safety)

Initials:_____   Date:_____

Issue	Action (Checklist)	Pass	Fail
Assessment of personal protective equipment (PPE)	Check that an assessment of PPE requirements has been made		
	Check that PPE conforms with relevant community provisions on design and manufacture. PPE carrying the CE mark will comply		
	Check that PPE is appropriate for the risks involved		
Maintenance and storage of PPE	Check that PPE examined and maintained on a periodic bases		
	Check for adequate stock of replacement PPE and PPE parts		
Training	Check for adequate and relevant training in the use of PPE and education in the risk of not using PPE		

Envirocompany Environment manual	SECTION: 2.3.0 PAGE 1 OF 1
Non-conformance	REV: 001 DATE: 16/6/92 Quality manager: J. Green

2.3.0 Non-conformance

Here we outline all the steps to be taken in cases of non-conformance. ISO 9000 procedures can be copied for these.

> The checklists, controls and monitoring logs suggested must be customised for each company to take account of the kinds of non-conformances which can be expected.

> This is an area of fundamental importance as far as inspections are concerned and will be the first that a company seeking accreditation to the standard will stand or fall on.

> Along ISO 9000 lines, the system should accommodate the following elements:
> Identification
> Segregation or isolation if possible
> Review
> Disposition if materials are available
> Documentation
> Prevention
> Corrective action

Envirocompany Environment manual	SECTION: 2.3.1 PAGE 1 OF 1
Normal procedures	REV: 001 DATE: 16/6/92 Quality manager: J. Green

2.3.1 Within limits

These are the reporting steps for all normal or within limits results.

The documentation already shown can be customised for each company and expanded as needed.

Envirocompany Environment manual	SECTION: 2.3.2 PAGE 1 OF 1
Incidents	REV: 001 DATE: 16/6/92 Quality manager: J. Green

2.3.2 Incidents

This will be specialized for each industry, but procedures and documentation for all outside of normal incidents must be maintained.

Envirocompany Environment manual	SECTION: 2.3.3 PAGE 1 OF 1
Emergencies	REV: 001 DATE: 16/6/92 Quality manager: J. Green

2.3.3 Emergencies

This is of such fundamental importance, with both corporate and personal risk of liability for management, that expert design and the best industry codes of practice should be used in both normal safety procedures and emergency procedures.

No further information is being offered here except that if the company does not have in-house or industry association expertise available, expert consultants should be used.

Envirocompany Environment manual	SECTION: 3.0.0 PAGE 1 OF 1
Audits and reviews	REV: 001 DATE: 16/6/92 Quality manager: J. Green

3.0.0 Audits and reviews

This section can contain the policy and procedures on auditing, especially where audits are in-house. It can also contain the records of audits with cross references to auditing procedures.

Envirocompany Environment manual	SECTION: 4.0.0 PAGE 1 OF 1
Training	REV: 001 DATE: 16/6/92 Quality manager: J. Green

4.0.0 Training

This section can contain records of staff training, such as logs. It can also contain rosters and be cross referenced to training manuals.

Index

Professional Report Writing

Simon Mort

<u>Professional Report Writing</u> is probably the most thorough treatment of this subject available, covering every aspect of an area often taken for granted. The author provides not just helpful analysis but also practical guidance on such topics as:

- deciding the format
- structuring a report
- stylistic pitfalls and how to avoid them
- making the most of illustrations
- ensuring a consistent layout

The theme throughout is fitness for purpose, and the text is enriched by a wide variety of examples drawn from the worlds of business, industry and government. The annotated bibliography includes a review of the leading dictionaries and reference books. Simon Mort's new book is destined to become an indispensable reference work for managers, civil servants, local government officers, consultants and professsionals of every kind.

Contents

Types and purposes of reports • Structure: introduction and body • Structure: conclusions and recommendations • Appendices and other attachments • Choosing words • Writing for non-technical readers • Style • Reviewing and editing • Summaries and concise writing • Visual illustrations • Preparing a report • Physical presentation • Appendix I Numbering systems • Appendix II Suggestions for further reading • Appendix III References • Index.

1992   232 pages   0 566 02712 7

Gower

Project Management
Fifth Edition
Dennis Lock

Project management is the function of evaluation, planning and controlling a project so that it is finished on time, to specification and within budget. It uses a family of techniques which can be practised with profit whether the project is worth £1000 or £1m.

Dennis Lock's book explains and demonstrates these techniques in action. It covers project management from initial appraisal to close-down, using methods ranging from simple charts to powerful computer systems, and with every subject explained using step-by-step illustrations and case studies. When it first appeared in 1968, it was acclaimed as a pioneering work: it is still the standard work on its subject, for managers and students alike.

This fifth edition has been completely revised and updated since the publication of the previous edition in 1988. Several new illustrations have been introduced and many of the existing ones replaced. Whilst the logical sequence of topics throughout the book remains the same the new edition now contains three additional chapters: on project management organization, on commercial management and on advanced procedures and systems. The result is an improved and strengthened edition that is completely up to date with the latest practice and technology. It is quite simply the standard text for anyone in industry who is interested in, or responsible for, projects.

Project Management is an Open University set text.

1992   542 pages   0 566 07339 0 Hardback   0 566 07340 4 Paperback

Gower

The Green Manager's Handbook

Kit Sadgrove

In many ways the environmental debate is over. For most organizations the question is no longer 'Shall we go green?' but 'How shall we go green?' <u>The Green Manager's Handbook</u> offers practical solutions, designed to help business people in organizations right across the business and public sector, incorporate green thinking into their policies and practices. <u>The Green Manager's Handbook</u> has valuable real life case histories, specimen policies, charts and procedures. It also has useful diagrams and dozens of checklists. Among the topics covered are:

- how to plan for a green future
- how to conduct an environmental audit
- how to reduce your energy bills
- how to develop and launch green products
- how to make your packaging more environmentally friendly
- how to make money from your waste products
- how to protect your organization from attack by pressure groups and conservationists
- how to create a business that people will want to work for
- how to apply environmental principles to your finance and accounting

Kit Sadgrove sees concern for the environment as a means to improved quality and efficiency. By cutting waste, using fewer toxic materials and producing environmentally sound products, companies can satisfy their customers and at the same time help to protect the planet.

1992   274 pages   0 566 07288 2

Gower

The People Side of
Project Management

Ralph L Kliem and Irwin S Ludin

This book explains the inter-relationships among the major parties of a project and provides ways for project managers to ensure cooperative, harmonious relationships.

It is written for everyone in business working on a project, regardless of industry. It addresses the psychological and political gaps that affect the outcome of projects. Senior management, project managers, project team members, and clients will all benefit from this book, particularly in mid-size and large firms where the "people factor" plays an important role. The book will enhance a better understanding of the "ins and outs" of how major participants of projects think, relate, act and interact.

It identifies the major players in a project environment and discusses the relationships (referred to as the people side) that exist among them from the perspective of the project manager. It discusses the impact of these relationships, throughout the project lifecycle, on major project activities, such as planning, budgeting, change management, and monitoring. It also discusses how project managers can improve these relationships: topics include leading individual team members, motivating the entire team, dealing with the client, and dealing with senior management. Finally, it discusses the qualities of effective project managers that engender cooperative, harmonious relationships among project participants.

1993   200 pages   0 566 07363 3

Gower

ISO 9000

Second Edition

Brian Rothery

This is a completely new and revised edition of the highly successful first edition of ISO 9000. This new edition takes account of all of the latest post-1993 changes and additions to the ISO 9000 series of standards, and looks ahead to emerging committee drafts and documents up to 1996.

Some highlights of this new edition are a completely revised generic Quality Manual for manufacturers, which can be taken and customised by any company in manufacturing and design; a brand new services Quality Manual, courtesy ICL, which should be a huge help to service industries adopting the standard; an update on the regulatory situation; and an exposition on how ISO 9000 fits in with other standards and regulations.

This book has been reflecting the worldwide spread of ISO 9000 itself, both in its sales, and in translations. Since first publication it has now become certain that ISO 9000 is the passport for trading, not just within the EC, but throughout the developed world.

The author has pitched the text so that it can be applied to quality managers and decision makers everywhere who are faced with the task of implementing or updating ISO 9000. This new edition will also be very useful for managers who intend also to implement BS 7750, the new environmental management standard, and who need a system to control the compulsory health and safety regulations, as these are cross-referenced to both of these issues.

Finally, the book provides new and penetrating insights into the questions of product liability and the increasingly mandatory nature of management standards such as ISO 9000.

1993   258 pages   0 566 07402 8

Gower

Building a Better Team
A handbook for
managers and facilitators

Peter Moxon

Team leadership and team development are central to the modern manager's ability to "achieve results through other people". Successful team building requires knowledge and skill, and the aim of this handbook is to provide both. Using a unique blend of concepts, practical guidance and exercises, the author explains both the why and the how of team development.

Drawing on his extensive experience as manager and consultant, Peter Moxon describes how groups develop, how trust and openness can be encouraged, and the likely problems overcome. As well as detailed advice on the planning and running of teambuilding programmes the book contains a series of activities, each one including all necessary instructions and support material.

Irrespective of the size or type of organization involved, Building a Better Team offers a practical, comprehensive guide to managers, facilitators and team leaders seeking improved performance.

Contents
Introduction • Part I: Teams and Teambuilding • Introduction • Teams and team effectiveness • Teambuilding • Summary • Part II: Designing and Running Teambuilding Programmes • Introduction • Diagnosis • Design and planning • Running the session • Follow-up • Part III: Teambuilding Tools and Techniques • Introduction • Diagnosis exercise • Getting started exercises • Improving team effectiveness exercises • Index.

1993   250 pages   0 566 07424 9

Gower

What Maastricht Means for Business

Opportunities and regulations in the EC Internal Market

Brian Rothery

A businessman's guide to the opportunities presented by the Maastricht Agreement, together with information on the implications of the resulting legislation for anyone managing a company in Europe, or trading with the European Community.

Contents

1993   200 pages   0 566 07430 3 Hardback   0 566 07431 1 Paperback

Gower